THE AI EMPIRE

How Hyperscalers, Compute, and the New Digital Colonialism Will Reshape the World

By

Basem Garada

COPYRIGHT

Paperback ISBN: 979-8-9949565-2-6

Published by
Digi Health Services LLC
United States of America

First Edition

DEDICATION

To every person who wonders what future their children will inherit.
And to those who still believe humanity can shape its own destiny.

CONTENTS

ACKNOWLEDGMENTS

This book could not have taken shape without the countless thinkers, researchers, technologists, historians, policymakers, engineers, and citizens whose questions, insights, warnings, and dreams helped illuminate the complex world we are entering.

I want to thank:

The scientists and engineers

who built the foundations of AI, compute, chip design, nuclear energy, and cloud infrastructure.
Your achievements make this era possible — and your caution reminds us of the risks we must confront.

The historians and philosophers

who taught that power has patterns, colonialism has logic, and technology always has a political dimension.

The policymakers and analysts

struggling to understand technologies that evolve faster than legislation.
Your efforts will shape global governance for generations.

The global researchers examining AI ethics, digital rights, and data sovereignty

who keep reminding the world that technology must serve humanity — not the other way around.

The open-source communities

who fight for a more decentralized and democratic future, often with limited recognition and fewer resources than their corporate counterparts.

The readers and critics

whose curiosity and skepticism fuel deeper exploration.
You are the reason this book exists.

And finally, to everyone who believes the future can be different

that empires are not destiny,
that technology can empower rather than dominate,
and that human agency matters even in the age of machines.

This book is for you.

EPIGRAPH

"Every empire begins as a promise of progress, and ends as a question of control."

INTRODUCTION - A New Empire Rising

We live in a moment when the surface of everyday life still looks familiar, yet the foundations beneath it are shifting faster than most people can see. Phones still light up, emails still arrive, search boxes still wait patiently for queries. But the intelligence animating these simple interactions is no longer a passive tool. It is becoming the engine of modern civilization.

Artificial intelligence now sits in the circuitry of almost everything that matters. It routes our messages, predicts our purchases, filters our news, prices our loans, translates our thoughts, watches for intruders, flags irregular heartbeats, and increasingly advises us on what to do next. Hospitals, banks, logistics networks, governments, armies, universities, and families are all, in different ways, learning to lean on machine cognition.

Yet beneath the excitement and the cascade of new applications lies a quieter and far more unsettling truth: the world's dependence is concentrating around a tiny handful of entities. A small cluster of hyperscalers—immense corporations with global reach—now control the great bulk of compute, the most capable AI models, the planetary cloud infrastructure, the energy-hungry data centers, and the technical priesthood required to run them.

In earlier eras, empires were built on coal and steel, on oil and shipping lanes, on manufacturing capacity and military force.

The great powers of the nineteenth and twentieth centuries counted their strength in tonnage of iron, barrels of crude, divisions of troops, miles of track and cable. Today, the hierarchy is written in different units: petaflops and exaflops, data center megawatts, access to advanced GPUs, sovereign or borrowed cloud regions, exclusive model weights, undersea cables and satellite constellations.

Compute capacity has become the new steel. Control of hyperscale data centers has replaced control of ports. Energy and cooling infrastructure underpin AI in the way railroads once underpinned the industrial age. Sovereignty is increasingly bounded not only by geography, but by who controls the platforms through which a nation's data, transactions, identities, and decisions flow.

These foundations of power are not evenly distributed, and they never will be. The capital and expertise required to build and operate global-scale AI systems are so immense that only a handful of actors can participate at the frontier. Amazon, Google, Microsoft, Meta, Apple, Nvidia, and a small circle of Asian giants now own much of the world's digital backbone. They control identity layers and authentication systems, the dominant email and messaging services, and the social networks where public life unfolds. They operate the clouds that host government services, the largest AI models and the chips that power them, the physical facilities that run day and night, and the satellites and fiber that stitch continents together.

Formally, they are corporations. Functionally, they are something more. They govern vast territories of human activity through contracts no one reads and interfaces everyone uses. Terms of service stand where colonial charters once stood. Algorithmic ranking replaces the censor and the editor. API access defines who may build on top of their worlds and who must remain outside. Moderation rules decide who may speak and who may be made invisible. Compute allocation quietly determines which projects, which regions, which languages, and which governments can participate in the frontier of AI at all.

This is power without borders, governance without geography, influence without elections.

There is another layer of this transformation that almost no one wants to talk about. Behind the screens and interfaces, behind the friendly chatbots and polished dashboards, the physical reality of AI is brutally simple. Models of growing size and appetite must be fed with electricity, cooled with water and air, anchored in land, and wrapped in regulatory permission. The more ambitious the systems become, the more demanding the installations that support them.

Already, data centers are competing with cities and industries for grid capacity. As AI workloads multiply, many regions are discovering that their existing energy systems cannot keep up. At the scale imagined for the next generation of models, renewables alone cannot provide reliable baseload. Fossil fuels collide with climate constraints and political resistance. Large nuclear plants face long approval cycles and fierce local opposition.

Into this gap steps a new logic: if the hyperscalers cannot easily expand their most intensive infrastructure at home, they will look abroad. They will search for nations with weaker regulatory environments, looser environmental constraints, cheaper land, and intense hunger for investment. There, they will offer a bargain—digital modernization in exchange for hosting the empire's engines. Land near coasts and deserts will become sites for AI campuses and cooling plants. In some cases, the power itself will be nuclear, supplied by small modular reactors built adjacent to data centers and tied more to corporate needs than to local grids.

The result is a modern version of a very old pattern: infrastructural colonialism. The environmental burden, radiological risk, and long-term land transformation remain with the host country. The strategic control of compute, the ownership of models, and the bulk of the profits remain with the hyperscaler. The periphery becomes a place where energy is transformed into cognition for the center.

All of this creates a global system that is both immensely powerful and disturbingly fragile. If one major hyperscaler were to fail, whether through financial collapse, cyberattack, sabotage, or geopolitical rupture, entire nations could find their hospitals, banks, logistics systems, and public services suddenly impaired. If export controls cut off access to high-end GPUs, a country's AI ecosystem could wither in months. If cloud access were throttled in a dispute, a government's own records might become unreachable.

Sovereignty in such a world cannot be measured solely in territory or troops. It must be measured in the ability to think, compute, decide, and store independently of someone else's platform. The question "who controls your borders?" is no longer enough. We must also ask, "who controls your compute?"

This book explores how we arrived at this threshold and where it may lead. It traces the mechanics of digital colonization, showing how compute became a new imperial weapon, how governments and hyperscalers together form a dual empire with overlapping interests and tensions, how certain regions are already drifting into the role of digital protectorates, and how AI's hunger for energy is pulling nuclear power and cloud infrastructure into an uneasy alliance.

It examines the early case studies between 2020 and 2025 that reveal the outlines of this emerging order: export controls that function as digital embargoes, Gulf states turning into global compute hubs, Africa and other regions sliding into a role as AI peripheries and data extraction zones, Europe struggling for sovereignty with strong regulations but weak infrastructure, China building a parallel technological civilization, and power grids bending under the strain of hyperscaler growth.

It follows the logic forward into a world where nuclear-backed AI campuses rise on foreign soil, where new forms of dependency and leverage emerge at the intersection of code and uranium, and where the future of humanity's autonomy is reshaped by a handful of entities that own both cognition and the reactors that power it.

This book does not claim that a dark, inevitable destiny has been written. Empires can be resisted, redirected, or dismantled. Nor does it suggest that AI is uniformly harmful; the same systems that can be used to dominate can also be used to educate, heal, and liberate. What it argues is simpler and more urgent: the trajectory is clear, unless it is confronted. The structures of a new empire are already under construction, poured in concrete and silicon, expressed in contracts and code.

We are building a new order, one in which intelligence itself becomes the primary terrain of power. The question that will define the coming decades is not whether this empire will exist, but who will govern it—and who will be governed by it.

PROLOGUE — The Invisible Empire

There are no emperors in this empire.
No crowns, no thrones, no royal bloodlines tracing power through generations. No banners hang from its buildings, no flags mark its territory, no borders appear on any map. And yet it governs more human lives simultaneously than any empire that has ever existed.

Its authority does not come from conquest, inheritance, or ideology. It comes from infrastructure. From code. From protocols written quietly in technical documentation. From data centers humming beyond public view. From cloud architectures and GPU clusters whose scale rivals the strategic stockpiles of past superpowers. From terms of service accepted without reading, and "safety guidelines" that decide what may be seen, said, generated, or remembered.

This empire is invisible because its borders are not drawn in land or sea. They are drawn in data flows. They follow packets across fiber-optic cables, through satellites, into server racks and memory banks. They appear wherever computation is required, wherever decision-making is outsourced, wherever a model interprets the world on our behalf before a human ever sees it.

It is present when a search engine answers a question before curiosity has fully formed. When an algorithm decides which news headlines rise and which disappear. When a credit system determines worthiness in milliseconds. When an AI assistant drafts words, suggests actions, predicts behavior, and subtly nudges choice. It is present not as force, but as mediation.

There was a time when many believed technology would decentralize power. That networks would dissolve hierarchies. That information abundance would weaken gatekeepers. That the internet would flatten authority and return agency to individuals and communities.

That belief did not survive contact with scale.

Instead of decentralization, we built concentration. Instead of many small centers of power, we created a few immense ones. Intelligence that once emerged from countless human minds is being funneled through increasingly centralized systems. Governments now run atop corporate cloud platforms. National infrastructures depend on privately owned compute. Digital identities—how we log in, authenticate, and exist online—are mediated by companies whose influence crosses every jurisdiction.

The paradox is striking: the systems that feel most personal are governed the least democratically. No population elected the architectures that shape their perception. No citizen referendum decided how algorithms should rank truth, relevance, or value. There was no treaty establishing who controls global AI. The systems simply appeared, expanded, optimized, scaled—and became indispensable.

The new empire has no Caesar, but its reach exceeds Rome.
It has no Senate, yet it influences legislation.
It has no army, yet nations depend on its infrastructure to function.

It does not march. It deploys.
It does not occupy land. It occupies cognition.

And unlike historical empires, escape is not a matter of geography. One cannot simply cross a border or sail to another shore. To participate in modern life is to pass through this empire's domains—to rely on its authentication, its computation, its models, its networks. Even resistance often requires using the very systems one might wish to question.

And yet, for all its reach, this empire is still forming. Its rules are not fully written. Its structures are not yet immutable. Its power, though immense, is not yet absolute. History is filled with moments when invisible systems hardened into permanent orders—moments when early awareness could have changed trajectories, but came too late.

We are living inside such a moment now.

This book is not written in the belief that catastrophe is inevitable, nor that technology is evil by nature. It is written from the recognition that power tends to concentrate when left unexamined, and that empires become most dangerous when they are invisible to those who live within them.

To understand the AI age, we must first learn to see the empire that is already here.
Only then can we decide what kind of world we are building—and whether we will remain its authors, or become its subjects.

BACK COVER SUMMARY

They said the future would be digital.
They never said it would be imperial.

As artificial intelligence accelerates beyond anything humanity has experienced before, a new world order is quietly assembling itself in plain sight. It is not built with soldiers or borders, flags or fleets. It is built with compute, cloud infrastructure, energy systems, and algorithms that mediate reality itself. At its center stands a small constellation of hyperscalers—corporations whose platforms now underpin governments, militaries, economies, healthcare systems, and the daily decision-making of billions.

The AI Empire reveals how this concentration of power is giving rise to a new form of colonialism—one that does not seize land, but dependency; that does not conquer territory, but cognition. It traces how access to GPUs and data centers has become a geopolitical weapon, how energy and nuclear power are fusing with AI to redraw global hierarchies, and how entire nations risk sliding into the role of digital protectorates without ever losing formal sovereignty.

Moving from real-world case studies to future projections, the book follows the emergence of competing AI blocs led by the United States, China, and an India–Gulf axis, while much of the world is relegated to the periphery. It shows how infrastructure decisions made today may lock humanity into a profoundly unequal technological future by mid-century.

Bold, unsettling, and deeply original, *The AI Empire* exposes the hidden power structures shaping the twenty-first century—and confronts the defining question of our time:

Who will control the nervous system of civilization?

CHAPTER 1 — When the World Quietly Changed

No one remembers the day it happened.

There was no announcement, no headline that captured it cleanly, no collective pause where humanity looked around and said, *this is different now*. People remember product launches, elections, pandemics, wars. They remember moments that arrive loudly. This did not arrive that way. It seeped in, thread by thread, update by update, until looking back became impossible because the before felt too distant to reconstruct accurately.

Life went on. People woke up, checked their phones, went to work, worried about bills, laughed at videos, navigated traffic, argued, dreamed. Yet beneath all of this, something fundamental had shifted. The world had begun to reorganize itself around invisible systems that most people never consciously acknowledged, but could no longer function without.

At first, it felt like progress. Things worked better. Tasks took less time. Answers came faster. Errors seemed rarer. Technology had always promised efficiency, and this looked like its fulfillment. Few questioned it, because questioning convenience rarely feels urgent.

For most of human history, power announced itself clearly. You could see it in fortresses, armies, factories, railroads, canals, and oil fields. If something dominated the world, it left marks on the landscape. Empires occupied space. Machines made noise. Exploitation was brutal enough to be visible.

This time was different.

The forces reshaping civilization were not loud. They were quiet, methodical, and distributed. They lived inside data centers most citizens would never visit, operating behind layers of abstraction carefully designed to make them feel neutral. Their influence was not exercised through orders, but through recommendations. Not through coercion, but through optimization.

No one told people to rely on them. People simply did.

The world had seen industrial revolutions before, and history books made them feel tidy. Steam reshaped economies. Electricity expanded reach. Computers collapsed distance. Each era disrupted labor, reordered geopolitics, and produced new forms of inequality. But each also left tangible evidence behind. You could walk through the ruins of old factories or trace power lines across continents and understand where the transformation had come from.

The current shift left no ruins. Only dependencies.

Somebody, somewhere, noticed this earlier than everyone else. Perhaps it was an engineer watching compute graphs climb beyond anything she had imagined when she entered the field. Perhaps it was a policymaker realizing that an entire national database now resided outside their legal jurisdiction. Perhaps it was a researcher who understood that without access to massive computation, certain ideas were no longer even expressible.

Or perhaps it was no one in particular — just a slow collective blindness caused by the fact that everything still worked.

Machines had always helped humans act on the world. They lifted heavier loads, moved faster goods, and amplified force. What was new was not that machines were becoming more powerful, but that they were beginning to mediate thought itself. They sorted information before humans saw it. They highlighted some patterns while burying others. They predicted outcomes and subtly nudged decisions long before consciousness entered the equation.

This did not feel like domination. It felt like assistance.

Yet there is a difference between using a tool and delegating agency. When a system decides what is relevant, what is likely, what is risky, and what is optimal, human choice becomes framed inside boundaries that are not always visible. Over time, those boundaries begin to feel like reality itself.

The world did not suddenly become artificial. It became curated.

This new intelligence did not come cheaply. It demanded resources older revolutions did not require. Vast amounts of computation, synchronized at a scale never before attempted. Energy not measured in kilowatts or megawatts, but in demands rivaling small cities. Cooling systems that needed rivers. Power contracts negotiated years in advance. Chips so specialized that controlling their supply meant controlling the pace of progress itself.

The raw materials of this age were no longer mined from the earth alone. They were manufactured through layers of industrial, logistical, and political coordination that few entities on the planet were capable of sustaining. The bottleneck was not creativity or ambition. It was access.

And access had begun to concentrate.

This concentration produced a new kind of actor in the world. One that did not fit existing categories. These entities were not states, but their reach exceeded that of many governments. They were not militaries, but they enabled military capabilities. They were not utilities, but entire societies depended on their uninterrupted operation. They spoke the language of markets and innovation, yet controlled infrastructure once considered national bedrock.

They did not conquer territory. They absorbed function.

Governments became clients. Institutions became tenants. Citizens became users. And no single decision ever marked this transition. It unfolded naturally, step by step, contract by contract, migration by migration, until reversing it no longer felt realistic.

What made this especially difficult to recognize was the lack of malice. Most of the people building these systems believed they were improving the world. And in many ways, they were. Disease detection became more accurate. Logistics became more efficient. Knowledge became more accessible. There was no singular villain to point to, no obvious line where innovation ended and dominance began.

But power does not require intent to exist.

The capacity to shape outcomes at scale is power regardless of motivation. When intelligence becomes centralized, its influence becomes structural. Societies begin to orbit around it the way ecosystems organize around a water source. Removing it would not simply cause inconvenience — it would cause collapse.

This reality revealed itself most clearly during moments of failure. A cloud outage here. An API disruption there. And suddenly airlines grounded flights. Hospitals delayed procedures. Payment systems froze. Public services stalled. The world did not panic for long, because the systems usually recovered quickly. But each incident exposed the same truth: resilience had given way to reliance.

Modern civilization had built its nervous system outside itself.

Another shift followed quietly behind the first — the return of energy as destiny. For a brief moment in history, it seemed as though the digital world had freed humanity from physical constraints. Data floated weightlessly. Knowledge traveled instantly. Geography felt increasingly irrelevant.

That illusion did not last.

Artificial intelligence, it turned out, was hungry. Every increase in capability demanded exponential growth in computation. Computation demanded energy. Energy demanded land, water, political concessions, long-term stability. Suddenly, where data centers were built began to matter as much as where factories once stood. Old questions returned under new names: Who provides the power? Who bears the environmental cost? Who controls access?

This reintroduced an uncomfortable symmetry with the past. Regions with cheap land and abundant energy became desirable hosts. Regulatory flexibility became leverage. Political alignment became a bargaining chip. What looked like investment often resembled extraction — except the resource being extracted was not mineral or labor, but digital sovereignty.

Not every nation could play this game. While public narratives spoke optimistically about democratizing AI, the material reality was less forgiving. The cost of entry climbed rapidly. Training state-of-the-art models required capital measured in billions, risk tolerance measured in decades, and infrastructure measured in physical footprints that could not be conjured by policy alone.

Talent mattered, but talent without compute stalled. Regulation mattered, but regulation without leverage bent outward. Vision mattered, but vision without resources waited.

The divide between those who could shape the future and those who could only consume it widened quietly and relentlessly.

This was not framed as exclusion. It was framed as partnership. As service. As opportunity. Nations were invited to migrate systems to the cloud. Enterprises were encouraged to modernize. Institutions were promised efficiency. And all of this was true — in the short term.

What was less discussed was the asymmetry it created.

When the intelligence layer of a society is externalized, autonomy erodes gradually. Decision-making becomes upstream-dependent. Customization depends on permission. Innovation becomes contingent. And eventually, even resistance requires access to the very systems being resisted.

This is how empires evolve when force becomes inefficient.

No borders are crossed. No laws are rewritten overnight. Control is embedded through architecture. Through defaults. Through the absence of alternatives. Through the quiet fact that building your own version now costs more than accepting someone else's.

The term "digital colonialism" unsettles people because it evokes history they would rather keep separate from the present. Colonialism feels crude, brutal, explicit. What is unfolding now feels clean, efficient, optional. But the structure is hauntingly familiar.

Resources flow from the periphery to the center. Infrastructure serves external interests first. Local agency diminishes without formal loss of sovereignty. And dependence is maintained not by violence, but by indispensability.

There are no governors. No flags. No visible chains. Only systems that work too well to question — until questioning becomes unavoidable.

This book does not begin by telling the reader what to think. It begins by acknowledging a sensation many already have and cannot quite name. A sense that power has moved somewhere difficult to see. A sense that technology is no longer merely a tool, but an environment. A sense that the rules of the world are being rewritten beneath the surface of daily life.

The Fourth Industrial Revolution did not arrive roaring. It arrived whispering. And by the time its presence became undeniable, it had already intertwined itself with everything society depended on.

The rest of this book follows the threads of that realization. Not to proclaim inevitability, but to illuminate structure. Not to assign blame, but to expose design. Because understanding how a system forms is the first step toward deciding whether it should continue as it is — or whether something else must be imagined before the quiet becomes permanent.

The world did not end when intelligence became infrastructure.

But it did change its rules.

And rules, once embedded deeply enough, stop asking for permission.

CHAPTER 2 — The Rise of the Hyperscalers

For most of human history, power had a recognizable shape.

It wore uniforms. It occupied land. It flew flags over buildings and carved borders into maps. Even when power was unjust, even when it was violent or exploitative, it announced itself clearly enough to be understood. You knew who ruled you. You knew where authority lived. You knew where to look when you wanted to resist.

The modern world inherited that intuition. We learned to see power in governments, armies, corporations, treaties. We assumed that influence scaled with territory, population, and visible force. And for a long time, that assumption still held.

Then something unfamiliar began to grow.

It did not seize land. It did not replace governments. It did not march or conquer. Instead, it positioned itself beneath everything else, quietly, efficiently, until entire systems could no longer function without it. By the time anyone thought to ask how much power it had accumulated, the answer had become uncomfortable.

These entities did not arrive calling themselves rulers. They arrived as service providers.

They offered storage, speed, reliability, optimization. They promised convenience, scalability, cost savings. They marketed themselves as neutral platforms rather than actors with interests.

They did not say they would become indispensable. They simply became so by solving problems no one else was prepared to solve at scale.

In the early days, it felt entirely harmless.

A company needed servers but did not want to build its own data center. A government wanted faster citizen services. A hospital needed better digital records. A university needed storage and computation. The cloud made sense. Outsourcing felt rational. Efficiency felt virtuous.

And somewhere along the way, the world crossed a threshold without noticing.

What emerged was not just a concentration of market power, but a concentration of infrastructure — the kind of infrastructure societies cannot easily survive without. These companies no longer merely competed with one another. They sat below competitors, below governments, below institutions, providing the invisible substrate upon which modern life rested.

This was the birth of the hyperscalers.

The term itself sounds technical, even boring. It hides more than it reveals. A hyperscaler is not simply a large technology firm. Size alone does not explain their influence. What distinguishes them is their ability to operate at a scale that collapses the boundary between the digital and the physical, between software and sovereignty.

They are capable of building computing systems that span continents and respond instantly. They deploy updates across billions of devices without pause. They store and process unimaginable volumes of data. They train models whose complexity exceeds the combined scientific output of entire nations. And they do all of this seamlessly enough that most users experience it only as normality.

What makes this power difficult to grasp is how mundane it feels from the outside.

You do not feel like you are interacting with an empire when you check email or upload a file. You do not sense geopolitical leverage when a document syncs smoothly across devices. Yet those small, quiet interactions are symptoms of a much larger consolidation occurring beneath the surface.

The rise of the hyperscalers was not accidental, and it was not inevitable. It followed a logic that rewarded foresight, patience, and relentless reinvestment.

Long before most governments understood the strategic importance of digital infrastructure, a handful of companies began building it anyway. They were not thinking in election cycles. They were not constrained by national budgets. They were incentivized to move faster, scale harder, and absorb losses others could not survive.

They built the digital equivalent of highways long before anyone realized that highways could determine economic destiny. Fiber networks sprawled beneath oceans. Data centers multiplied quietly at the edges of cities. Distributed computing systems evolved internally, optimized for scale rather than immediate profit.

At the same time, something subtler was taking shape.

Identity — once the exclusive domain of states — began to migrate into private systems. Logging in became an everyday act of governance, even if it did not feel like one. Authentication layers learned behaviors, locations, patterns, preferences. These systems did not merely verify who someone was; they learned how someone moved through the world.

Over time, logging in stopped being optional. Access to work, services, communication, and social participation flowed through corporate identity checkpoints. The border between citizen and user blurred.

This shift mattered more than it appeared to.

Identity is not just verification. It is control. It determines who can participate, who can be excluded, who can be flagged, who can be restored. When identity becomes privatized at scale, governance quietly migrates with it.

Alongside identity came data. Search queries, messages, photos, movement patterns, health signals, financial behaviors. At first, data was framed as exhaust — something generated incidentally. Then it became fuel. And finally, it became leverage.

AI did not emerge in a vacuum. It emerged inside ecosystems already saturated with information. The companies that had accumulated global data flows found themselves holding the raw material intelligence required. Others, no matter how brilliant, found themselves starved of the same input.

This created an asymmetry that compounded rapidly.

While most organizations were focused on quarterly performance, hyperscalers reinvested relentlessly. They poured revenue back into infrastructure, chips, research, hiring, energy contracts, and experimental systems that might not pay off for years. Their balance sheets began to resemble national budgets more than corporate finances.

For countries with limited GDP, the gap became impossible to close. Even if a government wanted to build competing infrastructure, the cost alone posed political risk. Failure at that scale did not mean a missed earnings call — it meant public backlash, budget crises, or leadership collapse.

The safer path was always to outsource.

This is how the cloud became the new empire.

Not by domination, but by absorption. Not by conquest, but by dependency. Governments migrated systems for rational reasons. Banks followed. Hospitals followed. Military logistics followed. Each migration was justified individually. Collectively, they reorganized global sovereignty.

At no point did anyone sign away independence consciously. They signed contracts. They optimized operations. They reduced costs. And slowly, control moved outward while responsibility remained local.

Once core systems lived elsewhere, the idea of reclaiming them became theoretical. Too expensive. Too risky. Too complex. Modern institutions do not pivot easily when infrastructure has been externalized.

Then artificial intelligence crossed its own threshold.

Suddenly, capability no longer scaled incrementally. It exploded. Model sizes grew faster than hardware roadmaps. Energy demands soared. Cooling requirements intensified. What had once been a software challenge became an industrial one.

Training a frontier model was no longer a matter of clever algorithms alone. It required entire campuses of machinery operating continuously, supplied by stable electrical grids and cooled by vast water resources. It required custom silicon, specialized networking, and teams of engineers whose expertise was limited to a small global elite.

Only hyperscalers could sustain this.

The moment AI reached industrial scale, the world's dependency deepened. It became clear that intelligence itself now passed through corporate chokepoints. Countries that wanted AI-driven healthcare, defense, logistics, or governance faced an uncomfortable reality: they did not own the means of intelligence production.

Even militaries — institutions historically obsessed with autonomy — quietly integrated hyperscaler infrastructure into simulations, logistics optimization, real-time analysis tools. It was framed as modernization. It was also a surrender of a certain kind of independence.

The relationship between governments and hyperscalers inverted.

Once, corporations depended on states for protection, regulation, and legitimacy. Now, states increasingly depend on corporate infrastructure to function at all. Hyperscalers negotiate directly with heads of state, influence legislation, and shape international norms. They attend summits not as stakeholders but as architects of what is feasible.

They do not overthrow governments. They make themselves unavoidable to them.

History offers an unsettling comparison.

In earlier centuries, chartered companies operated across the world with authority bordering on sovereignty. They extracted resources, shaped trade, enforced rules, and sometimes governed populations outright. They were privately owned yet state-backed, profit-driven yet geopolitically essential.

The difference today is not the logic — it is the medium.

Modern hyperscalers do not need armies. They do not need ships. They do not occupy land in visible ways. They embed themselves in systems so deeply that withdrawal becomes impossible without collapse. Their occupation is informational. Their influence is architectural.

Once a country's data, identity, intelligence, and infrastructure depend on external platforms, independence becomes performative rather than practical.

This power operates globally, but accountability does not.

Hyperscalers function across hundreds of jurisdictions simultaneously, yet answer primarily to boards and shareholders. Their rules are written in terms of service rather than constitutions. Their appeals are processed internally rather than in courts. Their influence reaches billions of people who have never consented meaningfully to their governance.

No one votes for them. Everyone lives under them.

As AI grows more demanding, hyperscalers are shedding their remaining illusion of immateriality. Their expansion is increasingly physical. Data centers rise like industrial forts. Power substations are built exclusively for compute campuses. Water usage reshapes local ecosystems. Energy contracts alter national planning.

In some regions, a single data center complex draws more power than surrounding towns combined. In others, water scarcity collides with cooling demands. Local resistance surfaces, not because people oppose technology, but because they feel its weight without controlling its direction.

This physicality drives hyperscalers outward.

Regulation tightens in traditional centers of power. Environmental scrutiny increases. Zoning becomes contentious. Energy costs rise. The solution is not retreat, but relocation. New regions offer land, energy, and political cooperation in exchange for investment and perceived modernization.

What they receive in return is less obvious.

Hosting hyperscale infrastructure creates dependency without ownership. The intelligence produced elsewhere flows outward. The economic benefits are uneven. Environmental costs remain local. Regulatory risk is assumed by hosts. And once embedded, the operators cannot be easily displaced.

It resembles outsourcing — except the function being outsourced is computation itself.

This creates a feedback loop that strengthens the hyperscalers year after year. More data enables better models. Better models attract more users. More users generate more data. Greater scale unlocks greater efficiency. And greater efficiency raises the cost of entry for everyone else.

By the time this loop becomes visible, it is already self-sustaining.

The world begins to divide quietly.

Not between democracies and autocracies. Not between rich and poor in traditional terms. But between those with sovereign access to compute and those without it. Between those who shape intelligence and those who consume it. Between those who negotiate and those who accept.

This is not a distant future. It is forming now, embedded in procurement decisions, infrastructure projects, trade agreements, and development strategies.

The hyperscalers did not plan to become empires in the classical sense. But empires are defined not by intent, but by structure. When control concentrates, when dependencies harden, when alternatives disappear, the label becomes less metaphorical than it sounds.

This chapter does not argue that hyperscalers are evil. It argues that they are powerful in ways history has seen before — and that pretending otherwise is dangerous. Power that goes unnamed goes unchallenged. Power that feels technical escapes political scrutiny.

The rise of the hyperscalers is not a footnote in technological history. It is the structural transformation upon which the next century will unfold. Diplomacy, security, culture, economics, and sovereignty will orbit around these entities whether societies acknowledge it or not.

The stage is now set.

What follows next is not about who these actors are, but how their power operates once embedded — how dependence becomes normalized, how choice narrows invisibly, and how digital colonialism functions without colonizers ever needing to arrive.

The hyperscalers have risen quietly.

And the world is already reorganizing itself around them.

CHAPTER 3 — The Architecture of Digital Colonialism

Colonialism used to be easy to spot.

It arrived on ships. It carried flags. It wrote its name on maps and carved its presence into stone and law. Its architecture was visible: forts, railroads, port cities, plantations, administrative buildings. The conquered could point to the structures that embodied their subordination.

The contemporary version is harder to see.

There are no uniforms. No parades. No declarations issued from palaces. There are only screens, contracts, networks, and code. Yet beneath those abstractions lies a structure every bit as organized as any old empire. It does not conquer territory. It colonizes infrastructure.

Most people never think of it as colonization at all. They experience it as digitization, modernization, and progress. A government moves its services online. A hospital replaces paper files with electronic records. A school adopts learning platforms. A city installs smart sensors to manage traffic, utilities, and public safety. The language surrounding these changes is relentlessly positive: efficiency, accessibility, innovation, transformation.

But somewhere between the first cloud migration and the thousandth API integration, something subtle happens. The technical backbone that runs a society stops belonging to that society.

Imagine a small country deciding to "modernize its state." Its leaders are tired of slow bureaucracy, crumbling servers, and lost records. Consultants arrive with polished presentations explaining how cloud platforms can host tax systems, identity registries, health records, and public services in a more secure, scalable, and cost-effective way. The argument is compelling and, in many respects, true. The country signs the contracts.

Within a few years, things look better. Citizens file taxes online. Hospitals pull up medical histories instantly. Births, marriages, property transfers—all of it lives in streamlined digital systems. The government is praised in international forums for embracing the future.

Then one day, a geopolitical crisis erupts.

Sanctions are considered. Diplomatic tensions escalate. Somewhere far away, in a boardroom the country's citizens will never see, risk assessments are updated. A hyperscaler marks this nation as a liability. A legal team revisits compliance obligations. A switch is flipped.

And in that moment, the country discovers that its digital nervous system does not live within its own borders, either technically or politically.

Much of the world now exists in this condition. The core databases of statehood—tax systems, identity registries, property records, health infrastructure, public finance, education platforms—are increasingly hosted on foreign cloud infrastructure, governed by foreign legal frameworks, and administered by staff answering to boards that owe no loyalty to the populations whose lives depend on their uptime.

This is the first layer of digital colonialism: cloud dependency that feels like modernization until it is tested.

Once the infrastructure moves outward, a second transformation begins, quieter still. The flows of data that pass through these systems do not simply sit where they are generated. They are mirrored, processed, aggregated, learned from. Behavioral patterns, linguistic nuances, cultural norms, commercial habits, social interactions—all of it feeds machine learning pipelines centered far away.

No one extracts gold anymore. They extract patterns.

Every search query, every click, every location ping, every interaction with a public service contributes to a growing portrait of a population. That portrait does not belong to the people who created it. It belongs to whoever has the computational capacity to capture and exploit it at scale. The resource is invisible, but the logic is familiar: value flows outward, benefit flows back only partially, and control over the resulting intelligence stays firmly in the hands of the extractor.

A country may believe it is merely renting servers. In practice, it is also exporting raw cultural material, behavioral signal, and linguistic richness, which are then used to improve foreign AI systems, foreign recommendation engines, and foreign language models. Those improved systems are then sold back into the world as superior products, further entrenching dependence.

The third layer appears when those AI systems become gatekeepers of meaning.

In past imperial arrangements, foreign powers reshaped local reality through schools, courts, churches, and administrative codes. They taught new languages, imposed new laws, and reframed what counted as legitimate knowledge. Today, the same work is increasingly done by ranking algorithms, language models, translation systems, and content moderation pipelines.

A student searching for information does not encounter a neutral universe of knowledge. They encounter the prioritized, filtered, and interpreted outputs of models trained mostly on data reflecting dominant cultures. A patient seeking health information is subtly nudged toward some explanations and away from others. A citizen sharing political views discovers that some expressions spread easily while others vanish into algorithmic obscurity.

These systems do not merely mirror culture. They gradually reshape it.

Local idioms get flattened. Minority languages struggle to fit into models optimized for high-resource tongues. Emotional expression begins to converge around what systems can recognize. Humor changes. Debate shifts. Even silence takes on new forms, because what is not legible to machines eventually becomes invisible in public discourse.

When a nation relies heavily on external AI for education, governance, healthcare, and communication, it also imports the worldviews baked into those systems. The result is not just technical dependency, but cultural dependency. A digital colonial architecture now extends into the realm of values.

The fourth layer is more mechanical but no less consequential. It concerns the veins and arteries of digital states: the application programming interfaces—the APIs—that connect systems together.

Every tax filing, health record lookup, payment authorization, identity verification, or logistics update flows through APIs. These interfaces are frequently controlled, rate-limited, priced, and updated by external providers. They are the gates through which modern life passes.

In theory, APIs are neutral technical constructs. In practice, they embody power. When access can be revoked, throttled, repriced, or conditioned on policy compliance, governance moves into the realm of interface management. A country's emergency response system could, in principle, be slowed down by a pricing change to a mapping API. Its businesses could be strained by new costs on payment gateways. Its civil society could be reshaped by changes to social media APIs that determine who can build tools for organizing or monitoring.

Control without physical occupation is possible when the infrastructure of connection is not under national authority.

Then there is the fifth layer, which rarely appears in political theory but dominates everyday digital life: the rule of contracts. Terms of service, license agreements, data-sharing policies, developer guidelines. These texts are not debated in parliaments or adjudicated in constitutional courts before they are applied. They are drafted by corporate legal teams and presented as "accept or leave" conditions to billions of people.

These terms define what content is allowed, which behaviors are punishable, what counts as abuse, who may participate, under what identity, and subject to which penalties. They govern whether encrypted communication can exist, how data is retained, which governments can demand access, and under what thresholds accounts can be disabled.

We live under these rules more intimately than under many national laws. Yet they are not anchored in democratic processes. They can be changed overnight and applied globally. The same clause can constrain a journalist in one country, a protester in another, and a student in a third, all without local input.

In this way, corporate contracts become a kind of transnational law—a legal superstructure that floats above nation-states, interpreting itself and enforcing itself algorithmically. The old slogan "no taxation without representation" slowly morphs into a quieter condition: no speech, no commerce, no identity without accepting terms written elsewhere.

The sixth layer arrives draped in an unassailable word: safety.

As AI systems become more powerful, companies face genuine pressure to prevent harm. They are urged to block misinformation, curb extremism, stop fraud, restrict incitement, and avoid enabling violence. Many of these goals are reasonable and necessary. The question is not whether safety matters, but who defines it, how it is enforced, and what else rides along with it.

Under the banner of safety, content is scanned, classified, ranked, and sometimes suppressed. Certain topics become harder to explore. Certain regions receive tailored output restrictions. Certain political or cultural narratives are downranked because they are deemed destabilizing, offensive, or risky.

What began as a shield against abuse easily becomes a tool for enforcing consensus.

The line between preventing harm and imposing ideology is thin, especially when billions of people have no visibility into how the line is drawn. An AI system that refuses to answer "unsafe" questions is not neutral; it is a mechanism of control. Its definition of unsafe, in turn, reflects choices made by institutions that are neither elected nor geographically bound to the populations they regulate.

Where colonial governors once wrote edicts, algorithms now enforce thresholds.

Finally, underlying all of this lies the seventh and most fundamental layer: the restriction of compute itself.

Even if a nation wished to escape these layers—to run its own models, host its own infrastructure, define its own cultural and policy boundaries—it would quickly discover that not all paths are open. Training state-of-the-art AI systems is not merely a matter of software. It demands the kind of computation only hyperscalers and a few great powers can marshal: vast GPU clusters, hardened data centers, dedicated power supplies, advanced networking, and budgets large enough to absorb failure without political collapse.

Countries without sovereign compute capacity find themselves constrained from the outset. They cannot easily train frontier models, cannot routinely retrain them with local data, cannot run large-scale experimentation on par with the global leaders. In practice, they must either import models as black boxes, rent access to inference through APIs, or content themselves with older, weaker systems.

The result is structural dependency.

Even if all other layers were somehow mitigated—if cloud contracts were fairer, if data-sharing were more reciprocal, if cultural representation improved, if API governance became more equitable—a simple reality would remain: the ability to produce high-end intelligence would still be concentrated in a small number of hands.

These seven layers—cloud dependency, data extraction, AI as cultural gatekeeper, API control, contractual law, safety-based filtering, and compute restriction—do not operate in isolation. They interlock, forming a lattice of constraints that is stronger than any single strand.

A nation might try to address one piece. It can negotiate better cloud terms, but still rely on foreign chips. It can store data locally, but still depend on foreign models to interpret that data. It can write its own content policies, but still live under the shadow of platform enforcement. It can dream of sovereign AI, but still lack the compute to build it.

In this interlocking system, modernization is both promise and trap.

On the one hand, no country wants to fall behind. Digital services genuinely improve lives. AI can enhance agriculture, healthcare, education, and logistics. Cloud platforms can reduce waste and corruption. To refuse these tools outright would be to accept stagnation.

On the other hand, embracing them without securing sovereignty means handing over the levers of self-governance bit by bit, often without realizing it. Each migration, each integration, each outsourcing decision is reasonable on its own. Yet collectively, they create a situation in which local political choices are constrained by external, unaccountable infrastructure.

The paradox is brutal: to modernize is to digitize; to digitize is to depend; to depend, under current conditions, is to be governed from elsewhere.

This architecture does not yet look like conflict. It looks like connectivity, partnership, progress. But the structure is already in place for future confrontations, coercions, and collapses. Power now has a blueprint that does not require armies or occupations. It requires only the right levers at the right points in the stack.

In the next phase of this story, those levers become explicit.

Compute, once a technical term in engineering conversations, begins to function openly as a weapon.

CHAPTER 4 — Compute as a Weapon of Empire

There was a time when the fate of nations could be read in coal shipments and oil contracts. The engine of empire ran on what could be burned. Steam and steel shaped the nineteenth century; petroleum and pipelines shaped the twentieth. Later, nuclear materials became symbols of both power and terror, compressing annihilation into the logic of deterrence.

Now, another resource sits quietly at the center of global strategy: not a fuel, not a mineral, not a radioactive core, but the ability to perform computation at an extreme scale.

For decades, compute was thought of as infrastructure, not as weaponry. It was racks of servers in corporate basements, mainframes humming in research labs, laptops on desks. Even supercomputers were mostly framed as tools for science. They predicted weather, simulated physics, modelled climate. They were prestigious, but not existential.

The arrival of frontier AI changed this calculus.

Training a state-of-the-art model is not like compiling an app. It is more like building, fueling, and running a factory capable of manufacturing intelligence itself. It requires specialized chips arranged into dense clusters, orchestrated by sophisticated networking, all sustained by continuous streams of electricity and cooled by industrial systems. The expense is measured not just in money, but in political capital and logistical complexity.

Very few actors can afford this.

Those that can have discovered something profound: control over high-end compute is control over what is possible in AI at all. Algorithms, data, and talent matter enormously—but without sufficient computational capacity, they cannot reach their full expression. The ceiling is not theoretical anymore; it is physical.

This is where computation begins to look like a strategic resource in its own right.

In previous eras, sanctions targeted banks, ports, shipping insurance, and energy trade. Today, an emerging form of pressure targets something that does not show up on docks or in barrels: access to advanced processors.

Denying a country oil slows its economy. Denying it high-end GPUs slows its future.

Without access to cutting-edge chips, a nation's AI research stalls. It can still build models, but only of a certain size. It can still innovate, but within narrower bounds. Militaries trying to modernize with AI-enhanced logistics, surveillance, or autonomous systems discover that their ambitions are capped by hardware they do not manufacture and cannot easily import.

Export controls that once applied to tanks or missiles now apply to boards of silicon. The gesture is quieter. The impact can be just as crippling.

This form of restriction is attractive to powerful states for several reasons. It is deniable, because it can be framed as non-proliferation or safety. It is selective, because specific models of chips can be targeted while others remain available. It is flexible, because laws can be adjusted faster than armies can move. And it is technologically enforced, because supply chains are traceable and manufacturers are few.

As a result, compute denial becomes a new kind of siege.

Instead of surrounding a city with soldiers, you surround a competitor's research institutions and industries with absence. There are no visible blockades, only procurement processes that never complete, shipments that never arrive, and domestic efforts that never fully catch up.

At the same time, on the other side of the equation, granting access to compute becomes an instrument of alignment.

When a hyperscaler or chip manufacturer decides which governments, universities, and companies receive priority allocation of scarce GPU clusters, it is not merely making a sales decision. It is shaping who gets to stand near the frontier of AI capability and who must remain in the wake.

Some countries are courted with generous partnerships: cloud credits, joint labs, shared research centers, favorable hosting arrangements. Others find themselves in a quieter queue, paying more for older hardware, with less say over terms. In diplomatic backchannels, access to infrastructure can become a sweetener, a reward, or a condition.

Nations no longer negotiate solely for bases, trade deals, or preferential tariffs. Increasingly, they negotiate for datacenter footprints, regional cluster deployments, and dedicated AI capacity.

The dual-use nature of compute amplifies this dynamic. The same clusters that generate images, summarize documents, and translate text can be repurposed to simulate battlefields, break encryption schemes, interpret satellite data, or optimize logistics for conflict. When a defense ministry signs a cloud contract, it is not just buying storage and virtual machines. It is intertwining its security apparatus with corporate infrastructure that may be subject to foreign policy considerations beyond its control.

This interdependence gives hyperscalers a peculiar position: they become part of the defense-industrial complex, but with reach beyond any single nation-state. They can effectively decide which military projects receive the most reliable access to advanced computation and which are limited. They can influence timelines, capabilities, and even doctrines indirectly through performance constraints and partnership choices.

For weaker states, this reality is harsh.

A country that cannot secure reliable high-end compute finds its intelligence services stuck deciphering smaller datasets while adversaries model entire theaters of operation. Its cyber defenses lag against automated attacks. Its weapons development suffers from slower simulation cycles. Its infrastructure—electric grids, transportation systems, ports—remains vulnerable because predictive models that could enhance resilience are too expensive to run at necessary scale.

The old measures of security—numbers of soldiers, size of arsenal, kilometers of border—become less decisive. What matters more is how much cognition a nation can wield in real time.

This stratification extends beyond defense. It reshapes global inequality itself.

At the top emerge a handful of compute aristocracies: countries and corporations whose combined access to chips, infrastructure, and expertise allows them to push the frontier. They define benchmarks, set standards, and sell services that everyone else must adopt or be left behind. Their companies become the primary providers of AI solutions in agriculture, finance, health, education, and governance.

Below them are mid-tier powers that can afford some degree of sovereignty—building national supercomputers, negotiating cloud regions, cultivating talent—but still rely heavily on external chip supply and software ecosystems. They have leverage, but limited control.

And at the bottom are those for whom AI is always imported.

These nations may sign impressive-sounding deals with hyperscalers, but they rarely own the means of production. They rent access to models through interfaces, purchase credits, adopt turnkey "AI platforms," and integrate them into their services. What they lack is the ability to independently shape these systems' core logic or to diverge meaningfully from provider roadmaps.

This is where compute inequality becomes structural.

It does not just mean some countries have fancier robots. It means some countries get to define what "intelligence" looks like in the digital age, while others adapt to it. It affects not only economic productivity but how problems are conceptualized, which risks are prioritized, and which solutions are deemed viable.

In this world, national security is inseparable from compute security.

Protecting borders without protecting access to computation is like defending a coastline while outsourcing the navy. You can still raise flags and march troops, but the most decisive maneuvers happen beyond your line of sight, in circuits elsewhere. Alliances become less about shared ideology and more about shared infrastructure.

Hyperscalers, for their part, increasingly find themselves behaving like geopolitical actors whether they intend to or not. When they commit to building a new region in a given country, they influence that country's strategic orientation. When they provide AI systems for identity verification, border control, policing, or social services, they shape domestic politics. When they participate in "public-private partnerships" for digital transformation, they sit at the same table as ministers and generals.

The boundaries blur. Are they vendors? Partners? Co-rulers?

Meanwhile, a small group of states attempts to escape this dependence by pursuing what they call sovereign compute. They invest in national chip fabrication facilities, supercomputing centers, homegrown AI frameworks, government clouds, and specialized education pipelines. They pour billions into catching up with pioneers who spent decades quietly constructing their advantage.

A few succeed partially. They create islands of autonomy—secure environments where critical workloads can run without reliance on foreign providers. They insist on local hosting for certain data. They negotiate hybrid architectures that keep some control close and push less sensitive workloads outward.

But even for these comparatively strong nations, the task is monumental. Chips remain deeply globalized in their supply chains. Toolchains depend on libraries and frameworks maintained by others. Talent flows toward higher-paying, more prestigious roles in global companies. And the frontier continues to move while they are still building their launchpads.

Most countries will never reach this threshold.

Not because they are incapable, but because the costs are simply too high and the time too short. The path of least resistance, again and again, runs through dependence: sign the deal, adopt the platform, trust the vendor, accept the terms.

This is why compute becomes the central driver of digital colonialism.

When a resource is this fundamental, scarce, and concentrated, it can shape the destinies of those who lack it. It can be withheld or granted. It can be time-limited, rate-limited, or price-limited. It can be framed as a business decision while functioning as political leverage.

It is not that tanks and missiles have become irrelevant. It is that the ability to design, simulate, coordinate, and deploy them efficiently now depends on invisible calculations happening in machines that may or may not be under a nation's control.

In previous chapters, we traced how digital infrastructure forms the skeleton of a new imperial architecture. Here, we see how one organ—compute—acts as both brain and weapon.

But compute does not float in the cloud in some abstract, immaterial sense. It is anchored in the physical world: in land, in cables, in substations, in generators, in reservoirs, and increasingly, in reactors.

To understand the next phase of this transformation, we must leave the pristine imagery of "the cloud" behind and walk out to where the data centers really are: on the edges of towns, by rivers, near power lines that glow at night.

There, the story becomes unmistakably material again.

CHAPTER 5 — Power, Cooling & Nuclear Energy as New Colonial Drivers

If you follow the fiber long enough, it eventually leads you to a building that does not care what the word "cloud" suggests.

It will not look like a palace. It might not even look sophisticated. From a distance, a hyperscale data center usually resembles a giant warehouse: a low, wide structure surrounded by fences, loading bays, satellite dishes, oddly placed ventilation fixtures, and an unusual amount of security.

Up close, you notice other details.

The hum is constant, a low mechanical roar that never stops. Trucks bring in equipment at odd hours. There are substations nearby, transformers big enough to power entire neighborhoods. Thick cables disappear underground. Cooling towers exhale invisible vapor into the air. Warning signs speak of high voltage, restricted access, and, in some cases, radiation.

This is where AI actually lives.

For a long time, the digital revolution masqueraded as weightless. We moved from discs to downloads, from paper to screens, from physical media to streaming. The metaphors we chose—cyberspace, virtual reality, online worlds—suggested a plane of existence separate from the material one. That illusion was fragile even then. Now it is collapsing under the sheer mass of what modern AI requires to function.

Teaching machines to approximate intelligence has turned out to be an extraordinarily energy-intensive project. Every new generation of models consumes more electricity than the last, not just once but repeatedly, as training, fine-tuning, and large-scale inference run continuously.

What used to be a server room in a corporate basement is now a 300-megawatt campus drawing as much power as a city of hundreds of thousands of people.

Grid operators feel this before the public does. They see industrial loads appearing in what used to be agricultural zones or light manufacturing districts. They receive connection requests that dwarf previous expectations: single projects asking for as much capacity as entire regions once used. Infrastructure timelines buckle. Long-term plans are rewritten. Politicians begin to ask whether the future of energy lies not in households or cars, but in buildings full of chips.

For hyperscalers, every watt is a constraint, every megawatt a bargaining chip. They need more power, more consistently, than most traditional industries. Unlike factories that can sometimes pause, training runs cannot easily be interrupted without losing progress. Uptime is not just preferable; it is essential.

This demand exposes another pressure point: cooling.

The waste heat from thousands of GPUs packed into dense racks cannot simply drift away. It must be moved. Traditionally, this has meant air conditioning on an industrial scale. As densities increased, systems turned to liquid cooling, evaporative towers, and more experimental approaches. Some designs consider submerged data centers in cold bodies of water; others explore new materials and architectural strategies.

For communities already worrying about droughts and water rights, the sight of a facility consuming millions of liters of water for cooling each day is hard to reconcile with narratives of progress. It feels like watching a thirsty machine drink from a shared well.

Protests emerge. Local councils hesitate to approve permits. Environmental organizations file challenges. The projects do not always stop, but they slow. Public opinion in many advanced economies begins to shift. Residents who once welcomed tech companies as clean, high-value industries now question whether the benefits justify the strain on land, water, and power.

Regulators respond. New rules cap water usage in sensitive areas. Environmental reviews become more stringent. Zoning laws tighten. Stricter emissions targets make carbon-heavy energy sources politically untenable. Large new data center proposals become harder to push through, especially when they come with the additional complication of AI's insatiable demands.

Faced with this resistance, hyperscalers look outward—to places where land is cheaper, regulations looser, and governments more eager for foreign investment. What follows is rarely framed as offloading risk. It is framed as development.

Negotiations begin with promises.

A country with high unemployment and crumbling infrastructure is offered the prospect of jobs, upgraded grids, digital services, and prestige. A gleaming "AI campus" becomes a symbol of modernization. Leaders imagine their cities on slides at global conferences, their flags represented in maps of the digital world's backbone.

What they are less prepared for is the depth of the commitment they are making.

A hyperscale data center is not just a building. It is a long-term claim on a region's energy budget. It is a perpetual draw on water. It is an anchor that dictates where new power plants must be built, which transmission lines must be reinforced, and how land use must be reorganized.

When these projects are paired with small modular reactors, the stakes rise further.

Nuclear power has always carried dual associations: immense promise and profound risk. For decades, public debates in wealthier countries have wrestled with questions of safety, waste, proliferation, and cost. Many societies have become cautious, even hostile, toward building new reactors in densely populated or politically sensitive areas.

But from the perspective of a company needing stable, carbon-light energy for computational loads that never sleep, nuclear's appeal is obvious. A single small modular reactor, properly deployed, can feed an AI campus for decades, insulated from grid volatility and price fluctuations.

If domestic resistance is high in the hyperscaler's home country, the logical conclusion—within current incentives—is to place the reactor elsewhere.

So, the offers evolve. Instead of just data centers, packages now include energy projects: private reactors, microgrids, integrated power-and-compute campuses. A foreign company arrives not only with a proposal to host AI infrastructure, but with a plan to reshape a country's entire electrical landscape in selected regions.

For the host nation, the pitch is tempting. Reliable energy, world-class infrastructure, an association with cutting-edge technology, and the infusion of capital all promise to lift some burdens. To say no may feel like turning down a future that will simply be offered to another country instead.

Yet buried in the fine print is a familiar pattern.

The nuclear plant may technically lie on the host's territory, but the intellectual property, operational control, and safety protocols often remain in foreign hands. The data center runs workloads not chosen by the host's citizens. The majority of profits flow outward. The primary beneficiaries of the compute are companies and partners elsewhere. What stays are jobs at the lower rungs of the value chain, increased pressure on local ecosystems, and an enduring dependency on technology and expertise that cannot be replaced easily.

This is not entirely new. The twentieth century saw similar arrangements with mining, oil, and heavy industry. The difference now is the resource being extracted and the intimacy of the infrastructure being installed. Land, water, and energy are still at stake—but they are now bound directly to the operation of systems that will govern information, identity, and intelligence.

Another dynamic emerges around water.

Regions with abundant freshwater sources but limited economic leverage become particularly attractive. Rivers fed by distant glaciers, deep aquifers, lakes that have sustained communities for generations—these are now also seen as potential cooling reservoirs for AI megacenters.

When the negotiation table is uneven, water becomes an asset traded cheaply.

A country struggling with debt may accept a deal that sacrifices long-term ecological stability for short-term cash and promises of digital uplift. Reservoir levels inch downward while data center capacity climbs upward. The immediate benefits are visible; the costs accumulate slowly and unevenly.

Over time, a new kind of state begins to take shape: the hosting protectorate.

Unlike historical protectorates, which were often explicitly under the "protection" of a foreign power, these new entities remain formally sovereign. Their leaders sit at international forums. Their flags fly at the United Nations. But underneath, critical parts of their territory operate as corporate enclaves.

Within these zones, private security forces patrol perimeters. Surveillance systems monitor gates. Access is controlled by badges, not by citizenship. Data stored inside may have more legal and technical protection in foreign courts than local residents have in their own.

The host government bears responsibility if something goes wrong—a leak, a meltdown, a cyberattack, a water crisis—but has limited power to dictate what happens day to day. As the number of such sites grows, so does this paradox: a country full of critical infrastructure that it does not truly control.

From orbit, satellite images will someday show clusters of these AI-energy complexes spread across the global south and parts of the global east: fortified rectangles etched into landscapes, hugging rivers, sitting on coasts, or tucked into sparsely populated highlands. Inside, algorithms trained elsewhere will churn through data from everywhere.

None of this arises because someone resurrected the word "colony" in a boardroom. It arises because the incentives line up perfectly.

Major powers and hyperscalers want more compute, more cheaply, with fewer domestic political complications and better climate optics. They do not want to shoulder the full environmental, social, and nuclear risk of feeding their AI habit at home. Weaker states want investment, legitimacy, and a foothold in the so-called knowledge economy. They cannot easily refuse offers wrapped in the language of partnership.

When both sides pursue their immediate interests, the long arc bends toward a familiar structure: extraction of resources from the periphery to support the core, external control over key assets, profits flowing outward, and burdens remaining local.

Only this time, the plantations are data centers, and the mines are aquifers and transmission corridors.

This is not to say that hosting such infrastructure brings no benefits. Some communities do see improved connectivity, new job opportunities, better roads, and upgraded grids. Some governments negotiate fairly strong terms, including local data requirements and profit-sharing. The picture is not uniformly bleak.

But the asymmetry is undeniable.

The ability to walk away lies overwhelmingly with the hyperscalers. They can relocate future investment, reroute workloads, and decommission facilities when they become outdated. The host country, on the other hand, is left with altered ecosystems, aging reactors, abandoned concrete shells, and tangled dependencies in its digital services.

The future, if nothing changes, is one in which the map of global inequality is traced not just by where wealth accumulates, but by where energy and water are sacrificed to sustain intelligence elsewhere.

Colonial logic has never cared much about the narratives used to justify it. At different times it has spoken of civilization, development, modernization, and partnership. In this century, it speaks of digital transformation, AI readiness, and climate-friendly innovation.

The underlying pattern remains: the powerful discover a new way to extend their reach without assuming the full cost. The less powerful accept the arrangement because the alternatives look worse.

The question the coming decades will answer is whether this pattern can be disrupted—whether it is possible to build an AI-powered world without recreating the extractive geometries of the past, only now with fiber and fission instead of sails and steam.

For now, the architecture is still being poured. The servers are still being racked. The coolant is still circulating. And as long as intelligence demands such immense physical support, someone will always be asked to carry the burden.

Who that someone is, and under what terms, is what the rest of this story must confront.

CHAPTER 6 — The Dual Empire: Governments and Hyperscalers

For most of history, empire wore one face.

It was the face of the state: kings and parliaments, ministries and councils, uniforms and seals. Corporations existed, and some of them became dangerously powerful, but they were always understood as subordinate in principle. They operated under charters, licenses, or concessions. Their excesses were framed as deviations from the proper order of things.

The twenty-first century has quietly overturned that arrangement.

Power still wears the face of the state, but beside it stands another figure, just as central, just as necessary, and in some ways less constrained. The two are not enemies. They are not even, properly speaking, allies. They are interlocked.

On one side stands government, carrying law, coercive authority, territorial sovereignty, and the symbols people still instinctively recognize as political. On the other side stand the hyperscalers, carrying compute, data, platforms, and the capacity to build, maintain, and direct the infrastructure of the digital world.

Separately, neither can fully govern the contemporary world. Together, they form something new: a dual empire.

This is not the product of a secret pact or a coordinated conspiracy. It is the outcome of scale, complexity, and necessity. The systems humanity has built are now so vast, so computationally intensive, and so entwined with daily life that neither the public sector nor the private sector can manage them alone.

The myth that still lingers, especially in media narratives, is that of "Big Tech versus Government." The story writes itself: rebellious founders clashing with regulators; hearings where CEOs are scolded; headlines warning that states are "finally cracking down" on technology giants; think pieces wondering whether democracy can restrain Silicon Valley.

There are real tensions. Fines are levied. Lawsuits are filed. Regulations are debated. But beneath the spectacle lies a more mundane truth: the two sides need each other far more than they openly admit.

For governments, hyperscalers are no longer optional vendors. They are the operating environment.

National tax systems are processed on cloud infrastructure. Social services are managed through hosted platforms. Voting systems, identity registries, immigration databases, licensing operations, land records—all of them increasingly sit on servers and software frameworks owned and operated by the same handful of corporations. Critical communications travel through their networks. Intelligence agencies rely on their storage, their compute, their AI models.

When a government announces it is "going digital," it is very rarely building that digital layer itself. It is outsourcing its nervous system.

This starts innocently enough. A ministry overwhelmed by legacy systems signs a contract with a cloud provider to modernize its operations. A health service, drowning in paper and incompatible databases, adopts a managed platform for electronic health records. A defense department seeks scalable compute for simulations and ends up anchored in a corporate cloud region. Each decision is rational at the level of the institution.

Add them together, and a state wakes up to find that large parts of its governance apparatus no longer reside within its own technical competence.

Officials log into dashboards they did not design, relying on code they do not own, running on machines they have never seen, administered by engineers who live under different legal and cultural frameworks. When outages occur, they do not call a subordinate agency; they open a support ticket.

Dependence becomes structural long before it becomes visible.

The dependency runs in the other direction as well.

Hyperscalers may appear self-sufficient, but their position rests on the scaffolding of state power. Intellectual property regimes protect their code and models. Trade agreements shield their cross-border operations. Export control frameworks, carefully tuned, are used to slow down foreign competitors while keeping their own expansion legal. Tax incentives, subsidies, and public-sector contracts supply revenue and legitimacy.

Governments, especially those of the most powerful states, serve as diplomatic fronts for their hyperscalers abroad. They pressure other countries to adopt compatible regulations, sign "digital partnership" agreements, open their markets, and host local data centers. In return, the companies align—at least broadly—with their home states' geopolitical strategies, declining certain deals, complying with specific restrictions, and sometimes acting as instruments of soft power.

The result is symbiosis rather than subordination.

The faces of this symbiosis are increasingly visible. When the head of a major cloud company visits a foreign capital, the choreography is almost indistinguishable from state diplomacy. Meetings with presidents and prime ministers. Closed-door sessions with ministers of finance, defense, and digital affairs. Memoranda of understanding. Joint press conferences describing partnerships in broad, optimistic language.

These trips are not mere sales tours. They are episodes of shadow foreign policy.

Questions once handled primarily between foreign ministries now pass through corporate negotiations: Who will host regional cloud infrastructure? Which identity systems will be adopted? Where will AI research centers be built? Which standards will define "trusted" technology? When a country announces a grand digital transformation plan, there is almost always a hyperscaler's logo attached, like a second coat of arms.

This interleaving extends inward into the machinery of governance.

Over time, governments have ceded not just hardware procurement and software licensing, but conceptual and operational control over their digital environments. Many agencies cannot fully describe the architectures that support their own missions. They know the names of the products they use, but not the details of how data flows, how models are updated, where the critical bottlenecks lie, or how failure cascades might unfold.

The deeper parts of these systems—the codebases, configuration layers, and machine learning pipelines—are protected as proprietary. Even when source code is shared or escrowed, the expertise required to meaningfully evaluate or modify it is often housed externally.

Digital governance, for all practical purposes, has become a managed service layer.

The stakes escalate once national security enters the picture.

As militaries integrate AI into their planning, logistics, surveillance, and targeting, the line between civilian cloud computing and warfighting infrastructure blurs. Hyperscalers provide simulation environments in which hypothetical conflicts are modeled. They supply tools for analyzing satellite imagery, detecting patterns in vast streams of signals intelligence, optimizing the movement of troops and equipment, and predicting cyber threats.

In previous eras, states developed such capabilities inside tightly controlled defense complexes. Today, many of the most advanced tools originate in commercial settings and are adapted for military use. The same cluster of GPUs that one week helps a company train a conversational model may the next week crunch data for a defense contract.

This creates an intelligence loop in which state and company share not just resources, but vantage points. Threats are jointly monitored. Vulnerabilities are jointly patched. Incidents are discussed in joint task forces staffed by government analysts and corporate engineers.

The military–industrial complex that defined the mid-twentieth century has not vanished. It has been layered with an AI–industrial complex, in which the same actors who control voice assistants and productivity suites also sit in hardened facilities discussing cyber intrusions and autonomous systems.

Where does public governance end and private capability begin?

The answer grows murkier when we consider how people encounter power in their daily lives. Between elections, legislative sessions, and court rulings, another layer of decision-making operates continuously and mostly invisibly: the layer of algorithms.

Search engines decide which information appears first when citizens look for answers. Recommendation systems determine which videos, articles, and posts surface in feeds. Moderation models and content filters determine which speech is amplified, which is suppressed, and which disappears without notification. Scoring systems in finance, insurance, hiring, and policing make judgments about who is trustworthy, who is risky, who deserves a second chance, and who is quietly excluded.

These systems feel personal and apolitical because they are embedded in consumer experiences: entertainment platforms, retail sites, messaging apps, news aggregators. But their cumulative effect is civilizational. They shape public opinion, economic opportunity, social norms, and even perceptions of reality.

Governments can complain about them. They can summon executives to hearings, propose rules, and sometimes force changes at the margins. But they do not operate these systems themselves. They do not own the training data. They cannot unilaterally override model behavior. Even when regulators demand transparency, what they receive is often selective and technical, not legible to ordinary citizens.

In this sense, algorithms constitute a fourth branch of governance: not legislative, not executive, not judicial, but allocative and interpretive. They assign visibility, attention, and risk in ways that profoundly impact lives. Yet they are written, tuned, and deployed by institutions whose direct accountability is to shareholders, not to voters.

This is where the dual empire reveals its deepest peculiarity.

On matters of law, physical coercion, and monetary issuance, states still hold formal primacy. They can arrest, prosecute, tax, conscript, and imprison. They can regulate, nationalize, and expropriate within their borders. But in matters of digital infrastructure, information flow, and artificial intelligence, hyperscalers hold operational primacy. They can block access, de-platform, de-prioritize, geo-fence, and alter the parameters of collective visibility.

Neither side can fully dominate the other.

States have discovered that breaking a major hyperscaler would risk catastrophic disruptions to their own systems. They can threaten antitrust actions, but the deeper their agencies embed into these platforms, the more they fear unintended consequences. Hyperscalers, in turn, have learned that open defiance of state authority invites regulatory and economic retaliation severe enough to derail their global ambitions.

The equilibrium that emerges is not one of harmony, nor of inevitable conflict. It is one of mutual constraint within shared dominance.

The dual empire operates by layering strengths.

Hyperscalers bring the infrastructure: the data centers, the networks, the models, the engineering talent, the product ecosystems. They also bring the intelligence—the literal capacity to generate, filter, and act upon insights at a scale no public bureaucracy can match. They shape the platforms within which public discourse and private life increasingly occur.

Governments bring the hard instruments: law, police, armies, taxation, diplomatic recognition, and control over officially sanctioned violence. They also bring the ability to define what is legal, who counts as a citizen, what contracts are enforceable, and which forms of economic activity are legitimate.

Together, they manufacture a form of power neither could create alone.

Consider a simple example: a system that identifies "high risk" individuals based on travel patterns, online activity, financial behavior, and social graphs. The model is trained, hosted, and maintained by a hyperscaler. It is run at scale only because the company has the necessary compute and expertise. But its outputs are acted upon by state agencies—border guards, police units, intelligence officers—who have the authority to detain, monitor, or restrict.

If citizens object, where do they direct their complaint? To the government that made the decision, or to the company that built the model that shaped the decision? To the parliament that funded the program, or to the board that approved the product? To a domestic court, or to a customer support portal?

Responsibility diffuses across the dual empire, even as its effectiveness concentrates.

None of this is happening by accident. Nor is it easily reversed. The complexity and cost of modern systems make it almost impossible for governments to "bring everything back in-house" or for hyperscalers to "stay out of politics." The path here is structural: as societies digitize, they create problems only combined public–private entities can solve. Once created, those entities learn to perpetuate themselves.

The open question is not whether this dual empire exists. It does. The open question is whether it can be governed—or whether it will, in practice, govern us.

In earlier chapters, we traced how digital infrastructure forms the skeleton of a new colonial order. Here, the outline sharpens. Standing atop that structure is a fused leadership class: elected officials who cannot deliver on their promises without corporate capability, and corporate executives who cannot fully secure their position without state backing.

They do not always agree. They quarrel, negotiate, threaten, and sometimes punish each other. But they operate within a shared architecture they both need too much to dismantle.

In this sense, the dual empire is not a glitch or a temporary phase. It is the backbone of twenty-first century geopolitics: a system in which sovereignty is no longer monopolized by states, but shared—unequally, uneasily, and irreversibly—with entities whose power was once mistaken for mere business success.

As the world moves deeper into the AI century, the shape of this shared sovereignty will determine much of what follows: how conflicts unfold, how resistance is possible, how law adapts, and how individuals navigate a world in which the ultimate answer to the question, "Who rules?" can no longer be given in a single word.

CHAPTER 7 — The Formation of Global AI Blocs

The world has divided itself many times.

Into empires and colonies. Into Christendom and the rest. Into East and West. Into capitalist and communist. Into "developed" and "developing." Each era found new lines to draw, new names for old hierarchies. People grew used to thinking in those terms. They knew what it meant for the globe to be split into ideological camps, economic zones, or military alliances.

Now, almost without language to describe it, another partition is forming.

It does not revolve around religion, or political systems, or even conventional wealth. It revolves around the ability to generate and deploy intelligence at planetary scale. The lines being drawn now are not simply between nations, but between clusters of nations bound to shared compute, shared clouds, shared chip supply chains, and shared AI models.

These are the AI blocs.

The peculiar thing is how quickly they are emerging. Oil took decades to reorder the world. Telecommunications and the internet took generations to be understood as strategic assets. AI has needed less than ten years to demand that countries choose where they stand, and with whom.

Part of the reason lies in how AI is built. It is not a distributed technology in the way early internet protocols were. It is not easily copied like industrial machinery once was. It depends on rare chips whose production is concentrated in very few places. It depends on hyperscale computing infrastructure that only a small number of actors can afford to construct. It relies on specialized talent, captured early and competitively compensated, clustering around particular ecosystems. And it is increasingly mediated by proprietary models that are offered as services rather than products.

All of this creates gravity.

When a country decides to "adopt AI," it finds itself pulled into orbits already defined by someone else's platforms, someone else's export controls, someone else's standards and roadmaps. Neutrality becomes technically difficult. Delay becomes economically dangerous. There is pressure—sometimes implicit, sometimes explicit—to align.

At the center of the most powerful bloc sits the United States.

Not because it alone possesses the technology, but because it anchors an ecosystem whose reach now extends across many of the world's wealthiest and most militarily capable states. Around it cluster Canada, the United Kingdom, Australia, Japan, South Korea, Singapore, and, increasingly, the Gulf monarchies that have decided to buy a seat near the frontier.

The strengths of this bloc are layered.

First, it holds the supply chokepoint: the design and distribution of the most advanced AI chips. Companies that design GPUs, and the fabs that produce them, are deeply embedded in this network. Export controls, licensing arrangements, and corporate strategies are interwoven. If this bloc decides a certain level of compute should not reach a particular destination, it usually does not.

Second, it owns the largest clouds. The platforms that now host a vast share of the world's workloads—the data centers, the global networks, the managed AI services—are headquartered here. Their regional footprints may seem distributed, but their legal and technical centers of gravity are not.

Third, many of the labs that are pushing the conceptual frontier of AI belong to this orbit. The models everyone else reads about, copies, or is forced to respond to—these are often born in a small cluster of companies on the western side of the Pacific and the Atlantic.

Fourth, the bloc is reinforced by longstanding defense alliances. Security arrangements originally built around tanks and ships and air power are now being infused with AI collaborations, shared research, and joint digital infrastructure. The idea of "interoperability," once a matter of compatible hardware and training doctrines, now applies to models, APIs, and data standards.

This combination gives the U.S.-led bloc a peculiar kind of authority. It can set the tone on AI safety, framing what counts as responsible development. It can shape export controls, determining who has access to what level of compute. It can propose architecture blueprints for cloud and data governance that others are pressured to adopt. Its diplomats do not travel alone; they carry the implicit backing of companies whose services everyone else has already integrated.

Yet for all its dominance, this is not the only center.

On the other side of the geopolitical mirror stands China, assembling a different kind of AI bloc.

China's approach cannot rely on the same components. It faces export restrictions on certain advanced chips. Its companies are often excluded from Western infrastructure deals. Its apps are viewed with suspicion or outright banned in some markets. But it has something else: scale, discipline, and a techno-political model that tightly integrates state power with corporate capability.

Over the past decade, China has built its own cloud platforms, its own AI companies, its own model stacks, its own data pipelines, and its own tightly controlled social platforms. It has poured resources into domestic chip research and production, sometimes lagging the frontier, sometimes closing the gap. It has turned its cities into testbeds for surveillance AI, smart policing, and integrated social credit mechanisms.

The appeal of this bloc is not universal. It is targeted.

Some states—those weary of Western conditionality, or deeply skeptical of Western values—are drawn to a model that promises digital modernization without the liberal rhetoric. They see that China can offer end-to-end systems: network infrastructure, cloud services, AI tools for governance, surveillance suites, financial platforms, all bundled in turnkey packages. They see that alignment with China's digital orbit may bring fewer lectures about human rights and more tolerance for centralized control.

Russia, increasingly shut out of Western technological ecosystems, gravitates toward this pole. So do some Central Asian republics, parts of Africa seeking infrastructure without Western scrutiny, and countries like Pakistan that straddle complex alliances. Iran, already under intense sanctions, perceives in Chinese support a way to keep developing digitally in spite of isolation.

The resulting bloc is different in character from the U.S.-led one. Where the Western bloc is a hybrid of powerful states and powerful corporations, with constant negotiation between them, the Chinese bloc is more overtly state-centric. Corporate actors exist, but their autonomy is constrained by political imperatives and party control. The architecture is not less sophisticated; it is differently governed.

Between these two gravitational centers lies Europe, awkwardly positioned.

The European Union recognizes the stakes. It speaks of digital sovereignty, of sovereign cloud, of sovereign AI. It passes some of the world's most ambitious data protection laws, crafts ethical guidelines, and pushes for transparent and accountable AI systems. It understands at a theoretical level that depending entirely on foreign hyperscalers for compute and models is risky.

But theory does not easily translate to hardware.

Europe does not own a hyperscaler of comparable size to the American giants. It does not dominate chip design. Its fabrication capacity is limited. Talented AI researchers born in European universities often end up working for U.S.-based companies. The continent's regulatory instincts, while admirable in their concern for rights, sometimes deter the very investment needed to build competing infrastructure.

The EU thus occupies a strange place in the emerging map. It is powerful as a regulator, able to shape privacy norms and require certain transparency measures that global companies grudgingly adopt. It can fine, restrict, and condition market access. It can slow down or redirect certain trends. But when it comes to the raw math of compute—the number of GPUs, the density of clusters, the budget for multi-year training runs—it lags behind.

Europe becomes, in effect, a moral and legal superpower in a world that increasingly rewards computational muscle. Its influence is real, but not decisive in the core contest over who trains the largest brains.

Elsewhere, a different configuration is taking shape.

In India and the Gulf states, a hybrid bloc is emerging that cannot yet rival the two main axes, but may in time become a third.

India brings scale of a different kind: a vast population, a deep pool of technical talent, and a large internal market hungry for AI-enabled services. Its government has historically tried to maintain a careful balance between major powers, buying weapons from one side while negotiating trade with another, participating in multiple constellations without fully committing to any. In AI, it appears to be pursuing a similar strategy: building national compute projects, exploring partnerships with Western and Eastern providers, and talking openly about sovereign data and models.

The Gulf monarchies, especially those swimming in oil and gas revenue, offer something India needs: energy, capital, and an appetite for grand projects.

Backed by sovereign wealth funds, they have begun buying vast quantities of GPUs, commissioning data centers in new mega-city developments, and announcing sovereign AI initiatives. Their ambition is not to be merely customers of foreign models, but to host and own their own, possibly in partnership with Western firms but under different terms.

Together, India and the Gulf are experimenting with a form of AI alignment that is not purely Western or Chinese. It is pragmatic, transactional, and anchored in a desire for autonomy without total isolation. If they succeed in building sustainable compute capacity, chip alliances, and regional AI ecosystems, they could anchor a bloc that offers smaller states an alternative: join neither the full Western orbit nor the Chinese one, but a looser network of countries that insist on some measure of technological self-determination.

Whether this experiment becomes a stable axis by 2040 is uncertain. It will depend on political continuity, sustained investment, and the ability to withstand pressure from the two larger blocs, both of which will see any independent center of AI gravity as a variable to influence.

Outside these zones of strength—and between them—are the majority of countries.

These states do not lack intelligence or ambition. They lack compute. They lack fabs. They lack domestic hyperscalers. They lack the surplus capital to spend billions on GPU clusters whose payoff may be uncertain and whose benefits often accrue after political time horizons.

For them, AI is not something they build; it is something they receive.

They sign deals with foreign cloud providers, adopt imported models, plug into pre-existing platforms. They gain capabilities they did not have before: automated translation, remote diagnostics, crop monitoring, educational tools. It is an upgrade, in many ways, and for governments struggling with limited resources, a welcome one.

But each integration deepens dependence. The code is foreign, the updates are foreign, the training regimes are foreign. If they want something tailored to local languages, customs, or needs, they must ask. If they want guarantees about data ownership, they must negotiate from a position of limited leverage. If they fall afoul of export controls or shift their foreign policy in ways their providers dislike, they may find access curtailed or priced differently.

These are the non-aligned in name, but the aligned in infrastructure.

They resemble, in digital form, the post-colonial Global South: formally independent, often rhetorically assertive about sovereignty, yet constrained by economic and technological structures they did not design and cannot easily change. Their "choice" of AI bloc is less ideological than opportunistic. They align with whomever offers the best terms at a given moment, or whoever is willing to invest in local connectivity, data centers, and skills.

They are, in effect, digital protectorates.

The real contest of the AI century may not be between the established blocs themselves, but over how these countries are drawn in, and by whom.

Africa, South Asia outside India, and Latin America—regions with large young populations, rich linguistic and cultural diversity, and enormous untapped data—are becoming prizes in a new kind of race. Not for territorial control, but for digital footholds.

Cloud regions are one kind of foothold. Telecom infrastructure is another. Educational partnerships that steer curricula toward particular platforms matter as well. So do AI projects in agriculture, health, and finance, which can lock institutions into specific ecosystems for decades.

The winner of this long campaign gains more than market share. They gain access to data that improves their models. They gain developers who grow up inside their tools. They gain friendly governments who shape local regulations in their favor. They gain political alignment in international forums where standards are set and norms agreed upon.

Unlike the Cold War, this struggle is not primarily about ideology. There is no grand story of capitalism versus communism animating citizens on both sides. The narrative frames that do exist—democracy versus authoritarianism, open versus closed systems—are sometimes deployed, but they lack the starkness of the past.

This is a competition of infrastructure and influence more than of manifestos.

By the time most people wake up to the fact that "their" institutions are deeply wired into one bloc's ecosystem or another, the choice will already have been made for them, through procurement decisions, pilot projects, and memorandum agreements signed years earlier.

From a distance, it might still be tempting to describe the world as unipolar, or bipolar, or simply chaotic. But those old categories miss what is structurally distinctive about the present moment.

We are not in a world dominated solely by one superpower, because no single country can monopolize all talent, data, and compute. We are not in a simple two-power confrontation, because regions like Europe, India, and the Gulf complicate the picture. We are not in a smoothly multipolar system either, because power is not evenly distributed among many.

We are in a compute-polar world.

The true axis around which global alignments form is not abstract "influence" or even pure GDP, but who stands closest to the frontier of AI capability and who controls the levers that determine access to it. Every alliance now has a digital clause: cloud sharing, AI research cooperation, chip supply guarantees, data reciprocity, cybersecurity commitments. Every trade agreement has, either explicitly or implicitly, a section about how information flows and where it is processed.

The blocs that emerge from this are not static. Countries will maneuver. Some will try to hedge between multiple poles. Others will test whether they can upgrade their status from customer to partner, from dependent to semi-sovereign.

But the underlying logic—that AI capability clusters into blocs defined by compute, chips, clouds, and models—is unlikely to fade. It is, like the dual empire of states and hyperscalers, a structural feature of the world we are building, not a negotiable preference.

The map of the twenty-first century will not be drawn only in colors of political ideology or lines of military alliance. It will be traced in latency, in the locations of data centers and fabs, in the opaque supply chains of chips, and in the legal jurisdictions that claim authority over models.

If previous centuries were defined by who controlled land and sea, this one may be defined by who controls the flows of intelligence.

And those flows, increasingly, run through the channels carved by the AI blocs that are, right now, still forming their shape.

CHAPTER 8 — Regions Most Vulnerable to Digital Colonization

Empires are rarely built where resistance is strongest.

They form where infrastructure is weakest, where urgency outruns capacity, and where the promise of modernization arrives faster than the ability to govern it. Digital colonialism follows the same logic. It does not target only the poorest or the weakest in a simplistic sense. It targets the least digitally sovereign—the places where intelligence can be imported more easily than it can be built, and where dependency is framed as progress.

Across much of the world, governments are not asking whether they should adopt AI, cloud platforms, or digital governance systems. That question has already been answered by economic pressure, global benchmarks, and public expectation. They are asking how quickly they can do so, and at what cost. The danger lies in the fact that the true cost is rarely visible upfront.

Vulnerability in this new era is not defined by GDP or military strength alone. It is defined by the absence of certain capacities. When a country cannot generate intelligence on its own terms, store and process its own data, or train models that reflect its languages and values, it is forced into a posture of dependence. It becomes a client not because it wants to be, but because there is no viable alternative path available at speed.

This is how protectorates are born—not through invasion, but through integration.

Africa sits at the center of this transformation.

No region combines such raw potential with such structural exposure. Africa is young, connected primarily through mobile technology, and increasingly urban. It generates enormous volumes of behavioral, linguistic, and geographic data, much of it in understudied languages and cultural contexts. From the perspective of AI developers, this is a frontier not yet fully mined.

At the same time, much of the continent lacks sovereign compute infrastructure. National data centers exist, but they cannot support large-scale model training or advanced analytics without external support. Power grids are often unstable. Cooling capacity is limited. Elite AI research ecosystems are rare and underfunded, and talent that emerges frequently leaves for opportunities elsewhere.

Hyperscalers recognize this imbalance clearly. What they see is a region ready for digital leapfrogging—but only through external platforms. Cloud-first government initiatives appear faster and cheaper than building national infrastructure from scratch. Digital identity systems sponsored by foreign firms promise inclusion and efficiency. AI tools for healthcare, agriculture, education, and public administration arrive pre-packaged, subsidized, and easy to deploy.

Each deployment tightens the weave.

Once government services depend on external cloud availability, once citizen identity is validated by foreign software stacks, once local developers learn only one dominant platform, exit becomes theoretical. Data flows outward. Models improve elsewhere. Local capacity remains shallow.

The irony is that Africa's richness—its diversity, its scale, its data—feeds systems that may never fully return that value locally. The resources extracted are not minerals or crops, but patterns of life. And those patterns increasingly shape global models trained far from where the data originated.

South Asia complicates this picture.

India, by virtue of its scale, talent pool, and political ambition, is actively resisting full digital colonization. It speaks openly about data sovereignty, national AI frameworks, and domestic compute initiatives. It negotiates with hyperscalers aggressively, extracting concessions and insisting on partial localization. It invests in talent pipelines and public digital infrastructure, seeking to remain a shaper rather than a customer.

But India is not the entirety of South Asia.

Pakistan, Bangladesh, Sri Lanka, Nepal—each faces different constraints. Their markets are large enough to be attractive, but their infrastructure is fragile. Their political situations fluctuate. Their fiscal space is limited. For them, hyperscaler-hosted systems are not only convenient; they are often the only way to meet immediate needs.

This creates a layered region where one country may attempt partial sovereignty while its neighbors become hosting zones, testing grounds, or consumption markets for foreign AI systems. The region becomes contested not through tanks or treaties, but through cloud contracts, telecom deals, and AI partnerships framed as development aid.

It is here that the competition between blocs becomes most visible. Western platforms promise openness and integration with global markets. Chinese systems offer turnkey governance without Western scrutiny. Gulf-backed initiatives dangle investment and compute access in exchange for alignment. South Asia becomes a digital battleground whose outcome will echo for generations.

Latin America occupies a quieter but no less precarious position.

Historically tied to U.S. economic systems, much of the region already runs on American technology stacks by default. Cloud infrastructure, enterprise software, social platforms, and payment systems are often imported wholesale. Brain drain is chronic; talented engineers move north. Local AI research struggles to compete for funding.

Yet the region also has strong universities, vibrant cultures, and growing internal markets. Brazil, in particular, shows signs of trying to carve out a more sovereign path, investing in national research and regional cooperation. The question is whether such efforts scale fast enough to counter gravitational pull.

Political volatility adds another layer of risk. Digital strategies change with administrations. Long-term infrastructure planning yields to short-term governance crises. Hyperscalers exploit this uncertainty by offering stability—platforms that persist even as governments turn over.

Dependency creeps in quietly. By the time risks are noticed, alternatives are expensive and politically difficult.

Southeast Asia may be the most strategically contested region of all.

Geographically positioned between major powers, economically dynamic, and culturally diverse, it attracts every bloc simultaneously. Different countries within the region lean in different directions, but none are immune to pressure.

Some align clearly with U.S. technology ecosystems, trusting Western companies more than Chinese ones. Others integrate deeply with Chinese surveillance and telecom systems, drawn by cost and speed. Still others attempt to balance both, hedging their bets in the hope of extracting maximum benefit while preserving autonomy.

The danger of this balancing act lies in becoming everything to everyone and sovereign to no one.

Social media platforms harvest data across borders. AI moderation systems influence speech in ways that local governments scarcely understand. Cloud providers compete to host services, locking institutions into incompatible ecosystems. The region risks becoming a kind of digital transit zone—a place where intelligence flows through, is refined elsewhere, and returns in prepackaged form.

Eastern Europe occupies a different kind of vulnerability.

Caught between regulatory dependence on the European Union, security dependence on NATO, and lingering influence from Russia, many countries in the region have limited agency in shaping their digital futures. Cloud adoption follows EU compliance frameworks, but the infrastructure is often foreign-owned. Local AI governance lags behind deployment.

In some cases, large data centers are built to serve Western Europe, not local needs. The region becomes an outsourcing zone for compute, hosting energy-intensive infrastructure without capturing the strategic benefits. Sovereignty exists on paper, but digital control is diffuse.

The outcome will depend heavily on broader geopolitical trajectories—on whether reconstruction efforts, especially in countries like Ukraine, include truly sovereign digital infrastructure or simply embed foreign systems more deeply.

In the non-Gulf Middle East, vulnerability deepens further.

Outside a few wealthy states capable of buying compute and talent outright, many countries face chronic instability, aging infrastructure, and limited fiscal space. They turn to AI-driven governance tools as a way to compensate for institutional weakness.

Surveillance systems promise order. Digital public services promise efficiency. Cloud-hosted identity systems promise control.

What is rarely discussed is how much influence these systems exert once embedded. Governance logic is imported along with software. Models trained on foreign data classify local populations. Cultural norms encoded elsewhere filter speech and behavior.

These states risk becoming corridors—routes through which hyperscaler megacenters spread, absorbing land and energy while exporting intelligence. The short-term gains are visible: jobs, investment, diplomatic goodwill. The long-term exchange—loss of autonomy—is harder to quantify until it is complete.

Small island states represent the extreme case.

They lack land, population, and redundancy. Climate change already threatens their physical sovereignty. Digitally, they are even more exposed. Entire government services rely on external platforms. Cloud deals promise resilience against disaster—but also lock in dependency.

Some become laboratories for new digital systems: experimental identity frameworks, AI-driven service delivery, regulatory arbitrage zones. What is framed as innovation is often necessity coupled with limited choice.

These are the first true digital colonies—not because anyone declared them so, but because survival itself depends on uninterrupted access to systems controlled elsewhere.

Across all these regions, the pattern repeats.

Legal sovereignty remains intact. Flags fly. Constitutions exist. Elections are held. But beneath that surface, digital sovereignty erodes. Decisions about infrastructure, intelligence, and data are made far from the populations affected by them. Governance becomes mediated through algorithms no one locally designed.

This gives rise to a new class of nation: the digital protectorate.

Protectorates do not surrender territory. They surrender autonomy over the systems that increasingly shape daily life. They negotiate terms rather than write them. They adapt to infrastructures rather than govern them. Their political discussions revolve around access, pricing, and compliance—not ownership.

Inequality in the AI century will not be measured primarily in income or life expectancy. It will be measured in compute per capita, in who can afford to train models rather than consume them, in who can set standards rather than follow them.

Countries that cannot close this gap—through energy investment, talent development, regulatory strength, and strategic restraint—risk falling into a permanent subordinate position. Not frozen in poverty, but locked into a dependent role within a global intelligence economy they cannot shape.

The tragedy is not inevitability.

Many of these regions possess the raw ingredients for a different future: human capital, cultural richness, strategic geography, renewable energy potential. What they lack is time. Digital colonization advances faster than traditional development cycles allow. Decisions made under pressure today define constraints for decades.

The next chapters will explore whether resistance is possible—what forms it can take, what trade-offs it demands, and whether true digital sovereignty can exist in a world where intelligence itself has become infrastructure.

For now, one reality stands clear: the regions most vulnerable to digital colonization are not those least capable of progress, but those forced to modernize fastest without the means to govern the tools doing the modernizing.

CHAPTER 9 — Can Any Nation Resist the New Colonial Order?

Most nations will live the AI century the way most people experience airline travel: seated, monitored, constrained by routes they did not choose, reliant on systems they do not control, grateful when the flight is smooth and helpless when turbulence arrives.

Only a few will ever sit in the cockpit.

This is not because most countries lack intelligence, ambition, or history. It is because resistance to the emerging digital empire is no longer a matter of rhetoric or policy statements. It is a matter of infrastructure, energy, talent, and time. And those ingredients are unevenly distributed.

The uncomfortable truth is that resistance is possible—but only under conditions so demanding that fewer than two dozen countries on Earth can realistically attempt it. Everyone else will face variants of accommodation: negotiating margins of autonomy inside systems built and controlled elsewhere.

To understand why, one must first abandon the comforting assumption that "good regulation" or "strong values" alone can preserve sovereignty. In the AI age, sovereignty is not primarily moral or legal. It is material.

It rests on whether a nation can generate intelligence without asking permission.

True resistance requires command over four overlapping dimensions of power. Remove even one, and the structure collapses into dependency.

The first is compute.

A country incapable of running large-scale data centers, operating sovereign GPU clusters, and training models at industrial scale is not autonomous. It can consume intelligence, but it cannot produce it independently. It cannot test alternatives. It cannot fully audit the systems it relies on. It cannot credibly threaten exit.

The second is data.

Without control over national datasets—health, education, mobility, commerce, language, culture—a country is feeding other people's models while impoverishing its own. Data sovereignty is not about isolation; it is about deciding where value is created. Once data is continually exported, the feedback loop works against the exporter forever.

The third is energy.

AI is not software in the old sense. It is an energy-hungry industrial process disguised as code. Without vast, stable, and politically controllable electricity supply, compute sovereignty is fiction. Models do not run on optimism; they run on megawatts.

The fourth is talent.

This is the hardest. GPUs can be purchased. Power plants can be built. Clouds can be negotiated. But elite AI talent—the mathematicians, chip architects, systems engineers, and theoretical researchers capable of pushing the frontier—cannot be imported at will. They take decades of investment, institutional credibility, and cultural gravity to accumulate.

Most nations lack all four simultaneously. Even wealthy states often master one or two at best.

This is why resistance is rare.

The most obvious path to resistance is brute force sovereignty: build everything yourself. Data centers, clouds, national models, GPU farms, and eventually domestic chip production. This is the path taken, in different forms, by the United States and China. It is now being pursued aggressively by a handful of smaller but extraordinarily well-capitalized states—particularly in the Gulf—and by advanced economies such as Japan, South Korea, and Singapore. France has begun tentatively. India is accelerating.

This path is brutally expensive.

It demands sustained investment measured not in millions or even billions, but in decades of strategic patience. It requires political consensus that survives election cycles, economic shocks, and public skepticism. It demands energy infrastructure capable of absorbing exponential growth in demand. It requires diplomatic strength to secure supply chains and withstand pressure.

Most importantly, it requires accepting that returns may not be immediate, visible, or popular. Building sovereign compute is not a product launch; it is a national gamble that pays off slowly and punishes hesitation mercilessly.

Nations that do not commit by the early 2030s will not catch up. By then, the frontier will have moved too far, the capital intensity too high, the ecosystem effects too entrenched.

For many countries, this path is simply closed.

This leads to the second viable form of resistance: regional alliances.

No single European state, for example, can rival American or Chinese compute on its own. Neither can most African, Latin American, or Southeast Asian nations. But coalitions can change the arithmetic.

Pooling resources allows countries to share GPU clusters, harmonize data governance, fund joint research institutes, and create sovereign cloud standards that dilute hyperscaler dominance. The European Union's efforts at cloud coordination and AI regulation reflect this logic, even if execution has been uneven. Similar experiments are emerging within ASEAN, the African Union, and Latin America. The India–Gulf axis represents an especially ambitious example of cross-regional partnership.

Regional resistance spreads risk and reduces individual vulnerability. It also creates shared bargaining power. A bloc that negotiates as a unit can extract better terms than a single state ever could.

But alliances are fragile.

They suffer from political fragmentation, uneven technical maturity, internal competition, and slow decision-making. They are vulnerable to divide-and-conquer tactics, where hyperscalers or major powers offer better bilateral deals to splinter the group. They struggle with crises that shift priorities inward.

For much of the Global South, however, this strategy represents the only realistic alternative to total dependence.

The third path to resistance is more modest, but often more viable: the "good enough AI" strategy.

This approach rejects the idea that every country must compete at the frontier. Instead of chasing trillion-parameter models trained on planetary-scale infrastructure, nations focus on sufficiency rather than supremacy.

They build open-source models fine-tuned on local languages and contexts. They deploy domain-specific AI for healthcare, agriculture, logistics, and public services. They prioritize edge computing and hybrid architectures where sensitive data never leaves national borders, even if some inference relies on external systems. They protect domestic innovation through targeted regulation and procurement policy.

This form of resistance does not produce global champions. It does not dominate standards-setting bodies or shape the global narrative of AI. But it preserves functional sovereignty. It allows nations to meet their citizens' needs without surrendering control entirely.

It is realistic. It is affordable. And it is incompatible with the fantasy that dependence can be resolved later, once systems are fully embedded.

What will not work is pretending that reliance on foreign hyperscaler infrastructure is temporary.

Digital dependence accelerates. Cloud systems are sticky. Data formats, APIs, authentication layers, and developer tooling entrench themselves deeply. Migrating away from them is not like switching vendors; it is like relocating a city.

Countries that base their public services, payments, identity systems, and governance platforms on external AI and cloud providers relinquish leverage the moment they sign. Every subsequent decision becomes constrained by sunk costs. What begins as pragmatic outsourcing becomes structural subordination.

Resistance must begin before dependency hardens. After that point, options narrow rapidly.

One truth cuts across every strategy: compute sovereignty requires energy sovereignty.

AI cannot be scaled on intermittent power, fragile grids, or politically unstable supply. Nations aiming to resist must confront uncomfortable energy questions. Nuclear power, geothermal systems, massive solar installations, hydropower, advanced grids, and aggressive cooling infrastructure are not optional luxuries. They are prerequisites.

This is one reason why certain countries punch above their demographic weight in AI ambition. Japan, South Korea, France, Singapore, and the Gulf monarchies possess not only capital but also stable, centralized energy strategies aligned with long-term planning. They can commit to infrastructure that will not pay off within a single political term.

Most nations cannot.

Energy scarcity becomes an invisible ceiling on sovereignty. And as climate constraints tighten, the competition for clean, high-density power will intensify.

Yet even energy and compute are insufficient on their own.

Resistance without cultural sovereignty is hollow.

AI systems do not merely process information. They shape language, normalize certain emotional expressions, filter political categories, and encode assumptions about identity and value. Models trained primarily on foreign data and aligned to foreign norms inevitably transmit those norms, subtly but persistently.

Without local models, local datasets, and cultural alignment work, nations face a slow erosion of narrative autonomy. Linguistic richness flattens. Minor languages wither digitally. Local ideas struggle to surface through global recommendation systems optimized elsewhere.

This is the least visible pillar of sovereignty, and often the most neglected. Cultural resilience requires investment in digital humanities, language institutes, archival digitization, and research labs that treat models not just as tools but as cultural artifacts.

A nation that controls its power and compute but loses its narrative voice will still find itself shaped from the outside.

Ultimately, however, all roads converge on the same bottleneck: talent.

Elite AI talent cannot be summoned into existence on demand. It emerges from ecosystems—world-class universities, research institutions, risk-tolerant funding environments, and cultures that reward technical excellence. It depends on immigration policy, academic freedom, intellectual prestige, and opportunities to work on problems that matter.

Nations can import hardware and infrastructure. They can subsidize reactors and data centers. But without domestic talent capable of understanding, challenging, and extending AI systems, those investments hollow out.

This is why talent mobility is increasingly politicized. Visas, research grants, and academic collaborations have become instruments of state strategy. Countries that fail to retain or attract top minds lose sovereignty by attrition.

When all these constraints are tallied, the list of plausible resistors shrinks dramatically.

At the top sit the high-probability sovereigns: the United States and China, followed by smaller but intensely focused states such as the UAE, Saudi Arabia, Singapore, South Korea, and Japan. These countries combine capital, energy, talent pipelines, and political will at scale.

Below them are partial sovereigns: France, India, Germany, Israel, Brazil. Each has strengths but also structural gaps. They may achieve selective autonomy but will remain entangled in global systems they do not fully control.

Everyone else—more than a hundred countries—will likely become digital protectorates to varying degrees. Not conquered, not erased, but constrained.

For the few that succeed, sovereignty in the AI age will reshape national identity itself.

They will treat compute like oil once was: a strategic resource to be secured, rationed, and defended. They will regulate clouds as critical infrastructure, not mere IT services. They will govern data as a national asset. They will embed AI readiness into defense planning, education, diplomacy, and energy policy. They will create ministries not just of technology, but of intelligence infrastructure.

For the rest, the question shifts from resistance to negotiation: how to preserve dignity, agency, and local value within systems they cannot redesign from scratch.

The AI century will not be kind to complacency. It will reward clarity, commitment, and courage measured not in speeches but in megawatts, models, and minds.

Resistance is possible—but only for those willing to pay its full price.

CHAPTER 10 — Lessons from Historical Colonialism

Every empire believes itself to be unprecedented.

At the height of its power, it insists that history offers no useful comparison, that its methods are novel, its motives complex, its structure modern. Those living inside such systems are encouraged to see them as the natural outcome of progress rather than a continuation of old patterns under new names.

History, however, has a different habit.

It does not repeat itself mechanically. It adapts. It changes instruments. It exchanges sails for steam, railways for fiber optics, and cannons for algorithms. Yet the underlying geometry of power—the way control is established, dependency enforced, and value extracted—remains stubbornly consistent.

Digital colonialism feels new because its tools are unfamiliar. But its logic is ancient.

To understand where this logic leads, it helps to look backward—not to indulge in nostalgia or moral grandstanding, but to recognize familiar structures wearing unfamiliar skins.

One of the clearest historical precedents emerges not from monarchies, but from corporations.

The British East India Company was not a state. It was a private enterprise, chartered for profit. It answered, in theory, to shareholders. In practice, it became a governing force over vast territories, wielding armies, levying taxes, shaping trade, and determining the fate of millions.

Its power rested on three pillars.

First, it controlled infrastructure: ports, shipping routes, and trading posts that others needed to access markets. Second, it enjoyed state backing: diplomatic protection, military reinforcement, and legal legitimacy when convenient. Third, it governed indirectly. It did not replace every local institution. It bent them, redirected them, hollowed them out, and rebuilt them to serve its interests.

No flag was planted at every injustice. No proclamation announced the end of sovereignty all at once. Control accumulated quietly, contract by contract, dependency by dependency.

If this sounds familiar, it should.

Hyperscalers are not armies, but they control routes—routes of computation, data, and intelligence. They are private entities, but they benefit from state protection, export regimes, and diplomatic muscle. They govern indirectly, through platforms, standards, and dependencies rather than through decrees.

Where the East India Company built ports along coasts, hyperscalers build data centers at network intersections. Where the Company controlled spice routes and textile flows, hyperscalers manage compute pipelines and data flows. Where Company officials "administered" regions under the guise of trade facilitation, hyperscalers administer digital territories under the language of optimization and efficiency.

The resemblance is not exact. It does not need to be. Structural similarities matter more than surface details.

Infrastructure has always been the silent hand of colonial power.

European empires did not merely conquer land; they rewired it. Railroads were laid not to connect colonies internally, but to extract resources outward. Telegraph lines enabled faster coordination between colonial capitals and metropoles. Ports and canals reorganized geography itself to favor imperial trade.

These infrastructures were sold as modernization. They were often genuinely transformative. But they were also irreversible. Once built, local economies, administrations, and cultures oriented themselves around them. Replacing them was prohibitively expensive. Operating without them was impossible.

Digital infrastructure functions the same way.

Cloud platforms, undersea cables, satellite networks, AI inference systems, and digital identity frameworks are today's railroads and telegraphs. They determine how value flows, where bottlenecks form, and who has priority access. They embed standards so deeply that alternative architectures appear impractical or "backward."

When a nation's hospitals, universities, banks, and ministries all rely on foreign cloud platforms, the infrastructure is no longer neutral. It is destiny.

The late nineteenth-century Scramble for Africa illustrates another instructive parallel.

European powers did not initially rush into the continent for ideological reasons. They were driven by competition: fear that rivals would secure land, resources, and strategic positions first. Conferences were held, borders drawn on maps, and influence carved out with little regard for local realities.

The real outcome was not simply territorial possession, but the creation of structural dependency. Economies were reshaped around export commodities. Governance systems were reorganized for control rather than representation. Local development became secondary to imperial extraction.

Today, a scramble is underway again—but the resource has changed.

The new contest is for data, hosting rights, energy access, model alignment, and digital identity control. Africa once again sits at the center of global competition, not because of weakness, but because of potential. Its youth, linguistic diversity, urban growth, and mobile connectivity make it invaluable to AI systems hungry for data and expansion.

But as before, the competition is asymmetric.

Powerful actors race to establish infrastructure first, knowing that early presence creates lasting lock-in. Agreements are signed quickly. Promises of development accompany platforms that shape dependency. What emerges is not overt domination, but an architecture that makes alternatives harder with each passing year.

The protectorate system of classical colonialism offers perhaps the closest mirror.

Protectorates were not colonies in the formal sense. They retained flags, rulers, and internal administration. What they surrendered—

often incrementally—was control over foreign policy, trade, security, and finance. The justification was protection: against rivals, instability, or internal disorder.

In return, the protectorate became dependent on the very power that claimed to shield it.

Digital protectorates function the same way.

A country may remain legally sovereign, culturally proud, and politically active. But its digital infrastructure—its clouds, identity systems, AI governance tools, and data flows—may be externally controlled. Decisions about uptime, security, pricing, content moderation, and access are made elsewhere.

There is no need for a resident governor. The platform governs.

Cultural influence, in colonial history, was seldom accidental.

Missionaries did not just spread religion; they reshaped language, education, moral categories, and social expectations. They brought scripts, grammar, schools, and norms that aligned colonized populations with imperial worldviews. The process was sometimes welcomed, sometimes resisted, often ambiguous—but its effects were long-lasting.

AI systems play a similar role today, not through intent, but through operation.

Language models prioritize dominant linguistic frameworks. Recommendation systems amplify certain narratives over others. Moderation algorithms define what is "acceptable" discourse. Emotional expression is standardized according to training data distributions. Political categories are framed in ways familiar to the systems' designers.

Over time, users adapt. Local idioms erode digitally. Minority languages struggle to find representation in large models. Cultural nuance is flattened to fit global standards optimized elsewhere.

AI becomes a kind of secular missionary—quietly reshaping how people speak, think, and imagine possibilities.

Resource extraction never disappeared; it evolved.

Rubber, gold, timber, and minerals once flowed outward from colonies to imperial cores, returning value upstream while leaving environmental and social costs behind. Today, the extracted resources are less tangible but no less valuable: attention, data, behavioral signals, cultural content, social graphs.

These raw materials are processed into models, products, and profits elsewhere. The originating societies receive services, convenience, and access—but rarely ownership or strategic control. The asymmetry persists.

Borders, too, have not vanished.

Colonial borders were imposed to facilitate control, divide populations, and optimize extraction. Digital borders now perform similar functions invisibly. API limits determine who can build and at what scale. Model access differs by region. Cloud service availability varies across jurisdictions. Content moderation rules shift across invisible lines.

Life on one side of a digital border is not the same as on the other.

The core lesson running through all these historical parallels is simple and unsettling: colonial systems escalate.

They do not stabilize naturally. They deepen because the incentives driving them reward greater control, tighter dependency, and expanded extraction. The East India Company did not stop once it dominated trade; it sought territory and governance. European empires did not halt once ports were secure; they pushed inland.

Digital colonialism is driven by the same pressures. As AI becomes more central to economic, cultural, and military life, those who control it will seek to secure their advantage. They will invest more heavily in infrastructure. They will tighten standards. They will integrate systems more deeply. They will justify expansion as safety, efficiency, or inevitability.

Without countervailing force, the empire grows.

The purpose of looking backward is not to claim that the present is identical to the past. It is to recognize warning signs. Colonialism did not become oppressive overnight. It became entrenched because early arrangements framed as temporary or technical hardened into permanent structures before resistance could organize effectively.

The lesson history offers is not one of despair, but of timing.

When dependency becomes total, resistance becomes symbolic.

Digital colonialism is still forming. Its borders are not fixed. Its hierarchies are not yet absolute. The future described in the next chapter—the world of 2050—is not inevitable. It is the trajectory if current patterns continue unchecked.

History tells us what happens when they do.

The question is whether this time, knowing the pattern, humanity chooses to interrupt it.

CHAPTER 11 — The World in 2050 If Nothing Changes

Futures rarely arrive with drama.

They arrive quietly, disguised as continuity. The year 2050 will not feel like a rupture to the people living in it. There will be headlines, certainly. New products, new crises, new promises. Children will be born who know no other world. Adults will adapt, as they always have, telling themselves that change was inevitable, that things are mostly fine, that this is simply how progress unfolds.

Only in retrospect will the shape of the transformation become obvious.

If current trajectories hold—if no structural corrections are made, no new forms of sovereignty established, no serious constraints imposed on the concentration of intelligence—the world of 2050 will be defined less by borders and flags than by access. Access to compute. Access to models. Access to energy. Access to the systems that now sit between humans and almost everything they do.

At the top of the global hierarchy will sit a small group of actors that no longer resemble ordinary nation-states in any meaningful way.

They will be known, informally, as the compute aristocracy.

This group will include the United States and China, not because of ideology, but because of sheer accumulated capacity. It will include energy-rich states that converted fossil or renewable abundance into intelligence infrastructure early and decisively. It will include tightly coordinated regional blocs that managed to pool talent, power, and capital into sustained AI ecosystems. And embedded among them will be a few hyperscalers whose reach exceeds that of most governments, whose models effectively function as global cognitive utilities.

These actors will not merely “use” AI. They will shape what AI is.

They will train the frontier models that others can only license. They will set technical standards that quietly become global norms. They will dictate which forms of intelligence are considered safe, efficient, ethical, or legitimate. Their military systems will be deeply algorithmic. Their economies will be optimized by predictive engines running at a scale no outsider can audit or replicate.

For them, the AI age will feel empowering.

For most of the world, it will feel like something that happens elsewhere.

By 2050, the concept of the digital protectorate will no longer sound alarmist. It will sound descriptive.

Large parts of Africa, South Asia outside a few hubs, Latin America, Eastern Europe, and many island nations will operate on external intelligence infrastructure as a matter of course. Their governments will still legislate, tax, and maintain public order. Their elections will still be held. Their leaders will still speak the language of sovereignty.

But beneath that surface, the systems that make modern governance possible—healthcare analytics, education platforms, financial risk assessment, infrastructure management, climate prediction, and identity verification—will be provided by foreign AI stacks.

These states will host data centers they do not control. They will export land, water, and energy to support training runs whose outputs primarily benefit upstream actors. They will negotiate access to increasingly essential services the way earlier generations negotiated access to oil or credit.

Their independence will be real, but narrow. Their room to maneuver will exist, but within corridors defined elsewhere.

Daily life, meanwhile, will be mediated by AI to a degree that would seem excessive today.

Healthcare decisions will be filtered through diagnostic systems that triage patients long before human doctors intervene. Educational pathways will be personalized by cognitive profiling tools that assign children to tracks early and adjust them continuously. Employment screening will be almost entirely automated, with AI systems ranking, rejecting, or flagging candidates at scale.

Infrastructure will be invisible but algorithmic. Traffic flows, energy use, water distribution, logistics networks—all optimized continuously by predictive systems. Food production and distribution will be increasingly automated, guided by AI-driven models of climate, soil, and demand.

Identity itself will become more formalized, more numerical, more legible to machines. Who you are, what risks you represent, what opportunities you qualify for—all will be assessed in real time by systems very few people understand.

For citizens of the compute aristocracy, many of these systems will feel empowering, even liberating. They will be customizable, responsive, performant. For those outside, they will feel prescriptive. You will not choose your tools so much as inherit them.

Energy-rich nations will rise in influence in ways that surprise observers focused only on traditional metrics of power.

Because compute runs on electricity, and frontier AI runs on vast amounts of it, countries that secured stable, scalable, politically controllable energy supplies will gain strategic leverage. Some will do this through nuclear deployment. Others through geothermal, hydroelectric, or massive solar infrastructure paired with storage.

These states will attract AI megacenters the way oil-producing nations once attracted refineries. They will negotiate hosting agreements that embed them deeply into global intelligence supply chains. Their land will matter not for agriculture or industry, but for cooling efficiency, grid stability, and geographic proximity to fiber routes.

Geopolitics will reorganize accordingly. Energy policy will be AI policy. Environmental debates will intertwine with questions of intelligence dominance.

Across parts of the world, entire zones will emerge whose primary function is to house computation.

These will resemble cities in scale, but not in character. Dense with data halls, cooling towers, substations, and secure perimeters, they will consume more power than previous generations of industrial complexes. Many will be nuclear-powered. Most will be built in regions where environmental resistance is limited, regulation flexible, and economic necessity acute.

They will be privately operated, often under long-term agreements that effectively place them beyond meaningful local political control. Security will be handled by private forces. Access will be restricted. Local economies will depend on them, even as strategic decisions are made elsewhere.

The resemblance to colonial outposts will be difficult to ignore—fortified nodes of strategic value embedded in peripheral regions, extracting resources and exporting value upstream.

Culturally, the world of 2050 will feel flatter.

Languages will persist, but fewer will thrive digitally. AI-mediated English—or a small set of globally dominant languages—will frame much of what people read, watch, and produce. Emotional norms will converge as models trained on global averages shape expression and response. Political discourse will be filtered through categories embedded in code, not necessarily grounded in local histories.

Creativity will not disappear, but it will increasingly reflect the aesthetic biases of the systems through which it is produced and distributed. Originality will coexist with subtle imitation, as neural networks trained on vast corpora guide what "works."

For nations that invested in cultural sovereignty—local models, language datasets, digital heritage—the erosion will be slower. For others, it will feel natural, unnoticed, and irreversible.

Within societies themselves, a new class divide will harden.

This divide will not be primarily economic in the traditional sense. It will be computational.

The compute elite will consist of those with privileged access to advanced AI systems: frontier models, personal AI agents tuned to their aspirations, educational environments optimized for rapid learning, networks that amplify their output. These individuals will think faster, learn faster, produce faster. They will compete globally from an early age.

The compute poor will live inside restricted AI ecosystems. Their models will be smaller, slower, more constrained. Their options will be pre-filtered. Their opportunities shaped by algorithmic pathways designed for efficiency rather than empowerment.

This divide will be self-reinforcing. Children born into compute-rich environments will inherit acceleration. Those born outside will inherit limitation. Formal equality will coexist with functional caste.

Warfare, too, will transform decisively.

By 2050, conflict will be dominated less by massed human forces than by autonomous systems facing other autonomous systems. Drone swarms, cyber offensives, AI-driven targeting, predictive logistics, satellite–AI fusion—these will define military balance. Speed, simulation, and adaptation will matter more than sheer numbers.

Nations without sovereign military AI will be effectively defenseless, reliant on allies or contractors for protection. Security guarantees will come bundled with compute access.

International politics will settle into an uneasy pattern.

Most countries will align formally or informally with one of the major AI blocs, not out of ideological affinity but because access demands it. Alignment will determine which models you can use, which chips you can buy, which clouds you can deploy, which standards you must follow.

This will not resemble the neat blocs of the Cold War. It will be messier, more technical, less visible to the public. But it will be no less consequential.

The world will not be unipolar, bipolar, or cleanly multipolar. It will be compute-polar. Power will radiate from centers of intelligence production, not merely from capitals or military bases.

And yet, even in this scenario, inevitability is a dangerous word.

History shows that empires appear permanent right up until they are not. Systems based on extreme dependency accumulate fragilities. Concentration invites shock. Monopolies provoke alternatives. Cultural flattening sparks revival movements. Energy systems break. Technologies diffuse in unexpected ways.

By 2050, decentralized AI, federated compute, new open-source paradigms, or global treaties may yet disrupt the picture described here. Movements for cultural sovereignty may gain traction. Environmental constraints may force rethinking of megacenters. Political pressure may impose limits on hyperscaler power.

None of these outcomes are guaranteed.

What *is* guaranteed is that without deliberate intervention, without early investment, without uncomfortable choices made soon rather than later, the world will drift toward the structure outlined in this chapter—not because anyone planned it, but because the incentives align and the resistance is weak.

The future is not fixed.

But it is being assembled right now, quietly, invisibly, at the speed of infrastructure.

CHAPTER 12 — Alternative Futures: Paths Away from Digital Empire

The future described so far in this book is not prophecy. It is inertia.

If no one interferes, if no major disruptions arrive, if states and societies continue making the easiest choices in front of them, then yes—the world of 2050 will likely resemble the compute-polar, hyperscaler-dominated order we have outlined. But history is full of moments when defaults were interrupted: not by perfection, but by enough friction, enough imagination, and enough resistance to bend the curve.

So the real question is not, "Is digital colonialism inevitable?" but "What has to happen for it not to be?"

Alternative futures do not arrive by magic. Each one has preconditions, trade-offs, and winners and losers of its own. None of them are cleanly "good" or "bad." But they matter because they show that the path ahead is not a single, pre-written script. There are forks in this road—even if some are narrow, steep, and hard to reach.

This chapter walks through six plausible alternative futures. Think of them not as predictions, but as map overlays. Each shows one way the landscape might evolve if particular forces gain strength. Some futures can coexist. Some will only appear in certain regions. All of them require work.

Every alternative future demands sacrifice—of convenience, speed, profit, or control—and none can be achieved without choosing what to give up.

The first path is the one many technologists quietly hope for, even if they rarely say it aloud: an open-source renaissance.

In this future, the gravitational pull of hyperscalers is weakened not by laws or revolutions, but by code. Open models—built by coalitions of universities, independent labs, national research institutes, and loose-knit communities—become good enough that they can stand beside, and sometimes above, proprietary systems.

For that to happen, several things must break in the right direction. Model compression and distillation would need to advance to the point where a system that once demanded thousands of GPUs can be run, in stripped-down form, on a modest cluster or even edge devices. Training methods would have to become radically more efficient, making it possible to fine-tune powerful models without access to industrial-scale compute. Decentralized compute protocols—using idle capacity across networks of machines, or local clusters stitched into federations—would have to move from theory to practice.

In such a world, a school district in Kenya, a hospital system in Peru, or a small-language media cooperative in Eastern Europe could deploy locally aligned AI on their own servers, behind their own firewalls, tuned to their own values. Governments would still negotiate with hyperscalers, but they would have credible alternatives. Citizens would know that not every intelligent system they touch is controlled by a handful of corporations in two or three countries.

For this renaissance to happen, open source cannot be a hobbyist afterthought. It needs real funding, legal protection, and infrastructure. Foundations and states would have to treat open AI as a public good, not a side project. Laws would have to defend the right to share and adapt models rather than shut it down under the banner of "safety." Hardware and software stacks would need to be intentionally designed to avoid lock-in.

If it happens, its disruptive potential is enormous. Not because it abolishes power, but because it redistributes leverage. Sovereignty becomes more attainable. Diversity of models—cultural, linguistic, ideological—becomes normal rather than exceptional. Hyperscalers do not vanish, but their monopoly over intelligence weakens.

The second future is more state-centered: a sovereign AI movement.

Here, governments react to the emerging digital empire not with resignation, but with national projects. They decide that relying completely on foreign clouds and foreign models for critical systems is intolerable. They build sovereign clouds, sovereign compute clusters, and sovereign foundation models. They pass data residency and governance laws that have teeth. They start thinking about AI alignment not as a generic question, but as something that should reflect their own legal systems, cultural histories, and risk tolerances.

This is not science fiction. You can already see its early outlines: Gulf countries pouring tens of billions of dollars into national AI ecosystems, India constructing a layered "national tech stack," France and a few others in Europe pushing for digital sovereignty even as they remain entangled with American infrastructure. In this future, those hints become full commitments.

Why would states choose this path? Fear and pride. Fear of vulnerability in a world where compute can be turned off cross-border. Pride in maintaining control over language, law, and culture. Economic self-interest, as leaders realize that whoever owns the models owns a large share of future value creation.

This future does not democratize AI globally. It creates pockets of autonomy. The nations that manage to build and sustain sovereign AI will enjoy margins of freedom others do not. They will still interoperate with global systems, but they will have bargaining power. They can say no, or at least "not like this."

The cost is high. Sovereign AI requires unprecedented state capacity: long-term planning that survives elections, deep investment in energy and infrastructure, talent pipelines that stop brain drain instead of feeding it. Many nations will try and fail. The ones that succeed will become a new kind of middle power: not necessarily the richest or largest, but strategically independent in the cognitive layer.

The third future shifts the focus from individual nations to coalitions: a multi-polar compute alliance.

In this scenario, regions that cannot achieve sovereignty alone decide to do it together. Europe moves beyond rhetoric and actually builds shared GPU pools, cross-border AI labs, and coordinated standards that mean something in practice. The African Union turns digital declarations into physical infrastructure, creating regional data centers governed collectively. ASEAN states manage enough trust to build common data frameworks. Latin American countries set up joint research centers to avoid duplicating scarce efforts.

The idea is simple: what is impossible for one country might be feasible for ten acting together.

Such coalitions become "NATOs for compute"—not military alliances, but mutual defense pacts against digital dependency. They negotiate as blocs with hyperscalers. They ensure that no single external actor can dictate terms unilaterally. They agree on baseline standards for privacy, data use, and model deployment.

The challenges are real. Regional mistrust, historical grievances, political turnover, and external pressures would constantly threaten to unravel such projects. Smaller nations might be tempted by side deals that undercut the alliance. But if they hold, these coalitions could create buffer zones against digital colonialism, particularly for mid-level states that would otherwise be easy targets.

The fourth future keeps hyperscalers at the center but changes their nature: the digital utilities model.

Imagine a world where the largest AI and cloud companies still exist, but are treated legally and politically more like power grids, railroads, or telecoms than like app developers. In this world, regulators impose price controls, interoperability requirements, enforced data portability, mandatory service for critical institutions, and structural separation between certain lines of business. National security rules restrict how these entities can act abroad. Supranational agreements cap their ability to hoard compute or unilaterally deny service.

Hyperscalers, in this scenario, do not disappear. They are domesticated.

This future would not be born of corporate enlightenment. It would come from a combination of public pressure, scandals, economic crises, and political leaders finally recognizing that letting a handful of firms govern critical infrastructure is too dangerous. It would require unusually strong regulators, legal creativity, and some degree of international coordination—because regulating a global actor with purely national tools is like trying to catch a whale with a fishing net.

If achieved, its disruptive potential is high. Not because it invents a new ecosystem, but because it changes how an existing one behaves. Instead of being accountable primarily to shareholders, hyperscalers would be forced to obey rules that reflect broader public interests. They would still hold enormous power, but they would no longer be free to wield it invisibly.

The fifth future is more speculative: breakthroughs in energy and compute efficiency that change the economics of AI entirely.

Present discussions assume that frontier AI will always require gargantuan clusters and staggering power. But imagine a sequence of breakthroughs in photonic computing, neuromorphic chips, or quantum accelerators that slashes the cost of high-end inference. Imagine ultra-efficient nuclear microreactors or continental geothermal grids that make clean, abundant electricity available far beyond current centers of wealth. Imagine design paradigms that allow high performance on cheap devices.

In that world, compute ceases to be scarce and hoardable. Intelligence becomes more like air than oil. The advantage of existing hyperscalers shrinks dramatically because their moat—industrial-scale infrastructure—is no longer a prerequisite for competitive AI.

This would not automatically produce justice or equality. Existing powers would still enjoy first-mover advantages. New forms of monopoly might appear. But the structural logic of digital colonialism would weaken, because the key resource would no longer be tightly bottlenecked. A rural university in a developing country could realistically run world-class models. A city government with modest means could host its own AI systems without begging for access.

The problem with this future is not desirability, but unpredictability. We cannot schedule breakthroughs. We can only increase their odds through public investment in research and insist that, if they arrive, they are not locked up instantly in restrictive patents and trade secrets.

The sixth future is the least technological and possibly the most surprising: a cultural rebellion against AI homogenization.

In this scenario, people begin to notice that their feeds look the same, their entertainments sound the same, their language is subtly converging toward global averages. They feel the edges of their cultures softening. They become uneasy with recommendation engines that know them better than they know themselves.

And they push back.

Not by abandoning technology entirely, but by making autonomy and local meaning fashionable again. Movements emerge around language preservation, digital minimalism, and culturally aligned

AI. Artists reject model-driven aesthetics in favor of deliberately local forms. Religious and philosophical communities demand systems that reflect their ethics rather than eroding them. Parents begin questioning whether handing children over to global AI tutors is wise.

States notice. Some support this resistance as a way to retain dignity. Others suppress it, fearing loss of control. The result is a patchwork world where in some places, AI becomes fully embedded, while in others it is kept deliberately at arm's length.

This future does not prevent digital colonialism everywhere. But it creates pockets where its advance is slowed or diverted. It reminds us that human beings are not purely consumers of utility; they also seek identity, rootedness, and difference.

No single one of these futures will dominate the entire planet.

Different regions will choose differently, or have choices made for them. Parts of the United States and China may double down on high-intensity techno-empires. India and the Gulf may embody sovereign AI futures. Europe may remain a regulatory power, shaping rules more than models. African countries may split between digital protectorates and coalition-driven resistors. Latin America may oscillate between dependency and regional assertion. Island states may lean into digital protectorate status simply to survive. Open-source movements may catch fire in some zones and remain marginal in others. Cultural rebellions may erupt in unexpected places.

The point is not to bet on one outcome. It is to recognize that the "default future" of consolidated digital empire is only one path, and that it becomes inevitable only if no serious alternative is nurtured.

The window to shape these alternatives is not infinite. Infrastructure hardens. Habits solidify. Contracts accumulate. But for now, the paths still exist. The question is not whether the world has choices, but whether it will use them.

CHAPTER 13 — The Logic of the Machine: Bias, Alignment, and Invisible Power

Up to this point, we have treated digital colonialism in largely material and institutional terms.

We have talked about data centers and GPU clusters, export controls and energy grids, treaties and trade-offs between governments and corporations. All of that matters. But it misses a quieter, deeper layer of power: the way the machines themselves think.

Empires do not only rule through ports and railroads. They also rule through categories—through the concepts that define what is normal, what is deviant, what is allowed, what is unimaginable. In the age of AI, those categories are increasingly encoded not in law or scripture, but in models.

That is the third layer of the digital empire: the cognitive layer.

Even if a nation someday succeeds in building its own data centers and hosting its own models on its own soil, it may still find itself governed by invisible logics inherited from systems trained elsewhere, aligned to values it did not choose, using data it did not control. This is where colonization slips from land and infrastructure into mind and meaning.

To see this clearly, you have to follow a subtle shift—from infrastructure to interpretation.

Traditional colonialism imposed obvious structures: new borders on maps, foreign legal codes, unfamiliar economic systems, racial hierarchies. The effects were brutal, but visible. Digital colonialism retains some of that, but adds something softer: it imposes filters.

Modern AI systems classify speech as safe or unsafe, relevant or irrelevant, trustworthy or suspicious. They rank which ideas appear at the top of your screen and which sink into statistical obscurity. They translate languages, but in doing so they favor some shades of meaning over others. They suggest which medical explanation is more plausible, which account of history is more "neutral," which answer feels more authoritative.

In thousands of quiet ways, they mediate reality.

When a government adopts a foreign AI model as the backbone of its education portal, or its legal information system, or its health guidance chatbot, it is not just outsourcing computing. It is outsourcing interpretation—outsourcing the way questions are framed, which answers are acceptable, which uncertainties are admitted.

These models are not neutral devices waiting to be filled with local meaning. They arrive preloaded.

We often talk about "bias" as if it were merely an unfortunate engineering flaw, like a bug to be patched. In practice, bias is the trace of human choices.

Consider what goes into a large model: decisions about which datasets to use and which to exclude, which languages to prioritize, which media outlets to treat as authoritative, which legal and cultural frameworks to anchor on. Even when teams try to be diverse and inclusive, they operate within constraints: they live in certain countries, read certain news, absorb certain norms.

The result is that models encode an implicit worldview. They decide, often unconsciously, which family structures are presented as default and which as deviations that require explanation, which political compromises are treated as obvious common sense and which as dangerous extremism, which religious expressions require careful distancing and which can be referenced casually, which dialects are tagged as "informal" and which are treated as proper language.

When these models are then deployed in societies whose histories, values, and conflicts look very different from those of the training data, something subtle happens.

Linguistic colonization reappears, this time through autocomplete and translation. Local idioms are "corrected" into smoother, globally familiar phrasing. Minority languages are handled poorly or ignored entirely, nudging speakers toward dominant tongues for anything digital.

Cultural colonization returns in softer clothes. Traditional practices are described, explained, or even pathologized through external moral frameworks: "rational," "modern," "progressive," on one side; "backward," "irrational," "superstitious" on the other.

Historical colonization persists in summaries and narratives. Local events are framed using references and priorities from foreign media archives. What one country regards as central trauma is reduced elsewhere to a footnote.

None of this requires malice. It follows from scale. When one set of cultures produces most of the digital corpus that trains most of the models, its perspective becomes the statistical default.

Alignment adds another layer of governance.

Publicly, alignment is framed as a safety problem: how to ensure models do not generate harmful content, how to prevent obvious abuses, how to keep systems "beneficial to humanity." Those are real concerns. But under that umbrella lie deeply political decisions.

Teams must decide what counts as "harm," which political claims are disinformation versus legitimate disagreement, which identities deserve special protection, which religious or cultural topics must be handled with extreme care, what sexual norms can be discussed, which wars or conflicts are "too sensitive" for detailed commentary.

These judgments are then encoded in guardrails, prompt templates, banned-topic lists, and fine-tuning datasets. The result is a kind of unwritten constitution for each model: a set of constraints that defines what the system will say, what it will avoid, and how it will frame controversial issues.

That constitution is not debated in parliaments. Citizens do not vote on it. Courts in most countries never see it. Yet it governs speech for billions of people who interact with these systems daily.

When a nation relies entirely on foreign-aligned models, it is effectively adopting someone else's unwritten constitution for speech and thought. Not formally, not ceremonially—but practically, which is what matters.

The story does not end there, because AI does not just mirror the world; it reshapes it and then learns from the reshaping.

Imagine a recommendation engine that favors a certain style of political commentary, or beauty standard, or romantic storyline. Users see more of that content. They engage with it, imitate it, produce more like it. That new content is then swept back into future training runs, reinforcing the system's sense that this is what people "want."

Alternative forms—more local, more idiosyncratic, more resistant—receive less exposure. They generate fewer clicks. They look rare in the behavioral data and then in the training data. The model learns that they are statistically unusual and that surfacing them frequently is "off pattern."

Over time, you get a feedback loop in which the model amplifies what it already privileges and attenuates what it sees as noise. Users adapt to what is most visible. The next generation grows up inside the output of the previous models. Diversity does not vanish, but it becomes harder to sustain.

In such an environment, AI is not just biased in the narrow sense. It is actively re-weighting the world that will train its successors.

Identity itself is not immune.

Where previous generations anchored selfhood primarily in family, religion, local community, and national history, the current and coming generations also anchor it in digital profiles: the playlists that know their moods, the platforms that suggest friendships, the AI assistants that help them think, the feeds that tell them what people "like them" are doing.

Models infer their beliefs, fears, vulnerabilities, and aspirations—sometimes more accurately than their friends or parents can. Then they shape the environments those people inhabit: which ideas they see, which communities they join, which role models appear, which products and careers are framed as "for them."

At first, this feels like helpful personalization. Over years, it begins to form a kind of soft corral. You can move, but mostly within a space drawn by algorithms.

If the AI systems doing this are built and governed entirely outside your culture, if they reflect alignment and bias choices optimized for distant legal and commercial environments, then the arena in which your identity takes shape is itself an imported product. A subtle colonialism of selfhood.

We usually talk about misalignment in catastrophic tones: superintelligences escaping control, machines optimizing away humanity. But there is another form of misalignment already with us: models aligned beautifully with the interests of their owners and misaligned with the interests of those who must live under them.

Platforms can design recommendation engines to maximize engagement, even if that drives polarization or addiction in certain societies. They can tune models to minimize legal risk in the jurisdiction that matters most to them, even if that distorts discourse in countries with different norms. They can choose cheap, one-size-fits-all deployment for poorer regions, even if that means those populations get cruder, less respectful systems.

From a corporate perspective, such systems are well aligned: they protect the firm, grow the metrics, and stay within the letter of key regulations. From the standpoint of national sovereignty and individual dignity, they may be deeply misaligned.

Opacity compounds the problem.

Frontier models are remarkably opaque by design: the composition of their training data is often secret; their architectures are summarized, not fully specified; their alignment protocols are presented in marketing language rather than in line-by-line detail. Even when companies release "model cards" or safety reports, these documents rarely answer the hardest questions: exactly whose content shaped the system's views, which governments or lobbying groups influenced sensitive-topic policies, which internal debates were resolved in which directions.

For countries that rely on these systems, opacity creates a dilemma. They are asked to integrate models into core governance functions—health, law, education—while being blocked from fully understanding how those models came to hold their apparent "knowledge." When a citizen is denied a loan, or a post is blocked, or a translation subtly changes political meaning, responsibility is diffuse: part code, part policy, part data, part accident.

Power without visibility has a particular flavor. It feels like inevitability.

On top of all this, AI is already becoming a regulator of behavior in its own right.

Content moderation tools decide which speech is visible and which vanishes. Fraud detection systems decide which transactions are allowed. Risk models used in banking, insurance, and housing quietly shape who can live where, under what terms.

Scoring systems in gig platforms determine who gets priority for work.

In many cases, these systems are technically supervised by humans, but at scale, the machine's judgment dominates. Appeals processes are opaque, slow, and often futile. Legislatures did not debate these decision rules. Courts did not test them against constitutional principles. Citizens never had a chance to vote on them.

For countries dependent on foreign platforms, a large share of everyday "regulation"—what people may say, do, buy, or build—is effectively outsourced. The line between corporate policy and social order blurs.

We often reassure ourselves by saying there is always a human in the loop. But we rarely ask which human.

Are they engineers in Silicon Valley, Shenzhen, or Dublin? Policy teams under pressure from American or European regulators? Contractors clicking through moderation queues at breakneck speed? Local experts from Nairobi, Jakarta, or Bogotá who can speak for the societies most affected?

In practice, the humans in the loop tend to share similar backgrounds. They work in a handful of institutions. They answer upward, not outward. Their incentives are to reduce corporate risk, not to defend the sovereignty of distant publics. They do their best, often under constraints that outsiders never see. But structurally, it is a narrow gate through which the experiences of billions must pass.

Classical colonialism had district commissioners writing rules for territories they barely understood. Digital colonialism has global trust-and-safety teams writing alignment policies for populations they may never meet.

If all of this sounds bleak, it is because we are finally looking at the right layer. But seeing the problem clearly is also the first step toward imagining something else.

What would non-colonial AI look like at the level of logic, not just infrastructure?

At minimum, it would require transparency about value choices: explicit documentation of what a model is optimized for, which harms it prioritizes avoiding, which legal and cultural frameworks informed its alignment. It would require the ability to localize alignment in meaningful ways, allowing regions and communities to adapt models within broad human rights constraints rather than being forced into a single global template.

It would demand genuine auditability: independent researchers, civil society groups, and governments able to test system behavior rigorously across demographics, languages, and contexts without being stonewalled by trade-secret claims. It would require mechanisms for contestation: ways for individuals and institutions to challenge AI-mediated decisions that affect their lives and to seek remedy when those decisions are wrong or unjust.

It would require a plurality of models. A world in which three or four closed systems mediate most digital thought is structurally colonial, no matter how those systems are draped in friendly branding. Sovereign AI, in the deep sense, means a landscape where many models—public, private, open, regional—compete and coexist.

And beneath it all, it would require something we rarely prioritize: algorithmic literacy. Populations that understand, at least in outline, that these systems are not oracles but engineered artifacts, with blind spots, power dynamics, and failure modes. Without that, people cannot resist, adapt, or demand better.

Data centers and GPUs matter. Export controls and nuclear reactors matter. But empires are sustained as much by the stories they tell and the categories they normalize as by their hardware.

The digital empire is no different.

If we limit our response to building national data centers while casually importing foreign logic, we will find that we have succeeded in reclaiming the land while leaving the mind colonized. We will own the metal and the megawatts, but not the meaning.

In the chapters ahead, we turn to the fragility of this empire—its environmental costs, its dependence on endless acceleration, the economic structures that sustain it, and the resistance movements trying to build alternatives. Because confronting the AI order is not just about challenging its towers of steel and silicon.

It is about challenging its invisible towers of thought: the architectures of logic that, if left unexamined, will quietly decide what humanity becomes.

CHAPTER 14 — FRAGILE FOUNDATIONS: ENVIRONMENTAL AND STRUCTURAL RISKS TO THE AI EMPIRE

The story of the AI empire is usually told in abstractions. We talk about "the cloud" as if it were a metaphysical realm floating above the planet. We talk about "compute" as if it were a magical fluid that can be poured anywhere at will. Marketing renders data centers as cool blue surfaces, softly glowing behind stylized security glass, untouched by weather, politics, or geology.

But none of that is real.

AI does not live in the cloud. It lives in buildings.

Those buildings are heavy. They sit on specific coordinates on the Earth's crust. They are made of steel, concrete, copper, rare metals, plastics, lithium, fans, pumps, and pipes. They gulp water and power. They shake when earthquakes hit. They heat up when heat waves linger. They hum, roar, and whine into the night. They provoke anger when they appear behind someone's backyard fence.

The digital empire looks invincible if you stare only at dashboards and market caps. Seen from the ground, it is surprisingly fragile.

This chapter is about that fragility. It is about deserts and ice, about noise and water, about floods and fire, about the simple fact that no matter how advanced the models become, they still run on hardware that must obey physics and coexist with human communities. It is a reminder that the same empire attempting to colonize the mental and political life of humanity is itself precariously perched on a restless planet.

The New Geography of Power

Data centers never appear by accident. They are not scattered like seeds by a careless hand. Each one is the product of a long calculation: where is the electricity cheap, the land abundant, the tax regime friendly, the political leadership eager, the climate forgiving, the fiber backbones close enough?

From those calculations, a new, strange geography of power emerges. It is not the geography of old empires—ports, coasts, and river mouths—but a geography of substations, cooling demands, zoning boards, and risk tolerances.

One of the most seductive regions in that geography is the desert.

In the Gulf, in places like Saudi Arabia and the United Arab Emirates, the sales pitch to hyperscalers is straightforward. There is land as far as the eye can see, away from residential neighborhoods, ready to be carved into artificial cities and industrial zones. There is energy, often subsidized and increasingly backed by ambitious renewable and nuclear plans. There is political continuity, with governments able to sign multi-decade agreements without fearing abrupt policy reversals. There is minimal public protest, at least in the traditional democratic sense.

On slides and glossy brochures, the desert looks like an empty canvas for the AI century.

But deserts are hostile to electronics. Heat pushes cooling systems to their limits. A data center whose racks already run hot must fight an ambient temperature that is sometimes closer to the inside of a server than the outside. Dust storms infiltrate any imperfection in sealing, layering fine particles on coils, filters, and fan blades, turning precision machinery into choking lungs. There is no cold air to pipe in during the night. There are no cold rivers to draw from. Everything must be generated mechanically, at enormous cost.

Most critically, deserts are thirsty. They do not have water to spare for lakes of condensate and evaporative towers. Every liter used to cool GPUs is part of a delicate, energy-intensive chain of desalination, pumping, and treatment. When a country with almost no freshwater commits to hosting dense clusters of AI infrastructure, it is also committing to ever greater entanglement between compute and energy, between energy and water, and between water and the basic survival of its people.

If the desert is risky because it offers too little water, the high latitudes seem appealing because they offer cold.

Nordic countries, and places like Iceland in particular, appear in strategy documents as environmental trophies: naturally chilled air, cheap hydropower, political stability, relatively small populations, and a green aura that appeals to both investors and regulators. You can almost imagine the diagrams: arrows showing cold winds flowing past tidy data halls, clean electricity flowing from turbines, all the messiness of AI neatly tucked into scenic landscapes.

But this “Nordic mirage” has its own weaknesses. Iceland sits atop an active volcanic system. The same geology that powers cheap geothermal plants also produces eruptions, ash clouds, and seismic events. Remote locations mean supply chains are long and maintenance teams are thinly stretched. Transmission capacity to export power is finite and contested. And even in these countries, citizens question whether their rivers and energy should be diverted to fuel global corporations' models rather than local needs.

Then there is suburban America, where much of the AI empire has quietly taken root. Here, hyperscalers gravitate toward flat farmland outside metropolitan regions, close to existing high-voltage lines and fiber trunks. They promise jobs, tax revenues, and prestige. What they bring instead, many residents discover, is a new kind of neighbor: faceless concrete boxes that never sleep, never quiet down, and consume electricity at a scale that dwarfs everything around them.

In these places, the geography of power collides directly with the geography of everyday life. The empire that wants to be invisible is suddenly heard through bedroom walls.

Noise: The Empire That Never Sleeps

If you have never lived near a data center, noise is an abstraction. You might imagine a faint hum, like a refrigerator. Residents in communities that host these facilities describe something very different.

They talk about a metallic roar that rises and falls but never disappears, a broadband drone that seeps into conversations, bedrooms, and dreams. They compare it to an airplane that started its engines and forgot to take off. They speak of windows that vibrate at night, of children who cannot sleep, of standing in their own gardens and feeling their chest cavity resonate with the low frequencies spilling from generator yards and fan walls.

The source of this soundscape is simple: data centers are giant life-support systems for silicon. Air must be moved, relentlessly, through racks and halls. Heat must be pushed out and away, in both normal operation and backup modes. Diesel generators fire up during tests or grid wobbles, adding their own bass rumble to the mix. Pumps, compressors, and transformers layer additional tones on top.

Unlike factories that shut down on weekends or at night, these machines never pause. The noise is the sound of uptime.

In town after town—outside Phoenix, in Dutch countryside, in Irish villages, in Virginia's now-famous "data center alley"—residents discovered that the hum of AI is not metaphorical. It is literally a hum, and it carries.

At first, the conflicts were treated as local nuisances: a zoning issue here, a lawsuit there. But as facilities multiplied, something larger became visible. Noise was not just an annoyance; it was a structural constraint. It slowed permit approvals. It triggered environmental review processes. It galvanized residents to organize, demand moratoriums, elect officials who promised to "stop the next data farm."

For hyperscalers, this created a dilemma. The easiest locations—suburban sites with good infrastructure and workforce access—were becoming politically expensive. To avoid the noise wars, they could push facilities farther out, into rural regions, deserts, or foreign countries where residents had less power to push back.

Noise, in other words, did not halt the AI empire. But it nudged it toward more fragile geographies, where environmental and political risks compound.

Water: The Hidden Thirst of Intelligence

Electricity often dominates the conversation about AI's physical footprint. But the empire's other addiction is water.

Modern high-density compute runs hot. Cooling it with air alone quickly runs into limits. Water, by contrast, is an excellent absorber and mover of heat. Whether it is used directly in liquid cooling systems or indirectly in evaporative towers and chillers, water is the silent partner in most large-scale data centers.

The volumes are staggering. A single hyperscale facility can consume millions of gallons of water per day, comparable to a town. A big AI training run can push that usage even higher, as racks are driven near their thermal limits for weeks on end. Much of this water is lost as vapor. Some can be recaptured, but not all.

In water-rich regions with modest populations, this might be tolerable. In arid zones, it is a different story.

The paradox is especially sharp in places like Arizona or the Gulf, where AI ambitions are rising fastest. Freshwater reserves are limited. Rivers are shrinking. Aquifers are stressed. To cool data centers, these regions turn to desalination, turning seawater into usable water at the price of huge energy input and concentrated brine waste.

The chain is easy to trace: more AI means more data centers; more data centers means more cooling; more cooling means more water; more water in desert regions means more desalination; more desalination means more energy and more environmental externalities. At the end of this chain stands a human question: when farms, homes, and ecosystems compete with GPUs for water, who wins?

In many parts of the world, the political conflicts of the 2040s are likely to be framed not as "AI versus humanity," but as "cooling versus drinking," "training runs versus rivers." The empire's thirst becomes its Achilles' heel.

Disasters: When Nature Ignores Uptime Guarantees

Marketing images of cloud infrastructure often show tranquil blue skies above neat rows of racks. The planet has other ideas.

The AI empire is already woven through regions that tremble, flood, and burn. California, home to countless corporate headquarters and major facilities, sits on fault lines and faces seasonal wildfires. Japan, with its dense tech ecosystem, lives with earthquakes and tsunamis. Turkey and Chile—key nodes in global connectivity—are likewise seismically active. Coastal regions in Europe, North America, and Asia endure stronger storms and rising seas.

A moderate earthquake may not topple a building, but it can shake server racks loose, damage raised floors, crack coolant pipes, and disrupt power equipment. Fire does not need to engulf a data center directly to cause havoc; thick smoke can clog filters, and emergency shutdowns can ripple through entire regions. Floods can render underground cabling useless, shut down substations, and isolate facilities that were built in what used to be considered "safe" lowlands.

Companies tout redundancy and "five nines" of uptime, and to some extent they deliver: systems are designed to fail over between regions, to reroute traffic, to shed load gracefully. But redundancy is not witchcraft. Switching active workloads between sites is nontrivial. It involves network reconfiguration, state synchronization, cache invalidation, database replication, and a careful choreography of traffic.

Disaster recovery plans look elegant in diagrams. In practice, they involve uneven human reactions, partial information, cascading small mistakes. What was supposed to be a seamless failover can turn into hours of partial outages. Days later, reports quietly note that “some users experienced degraded performance.”

There is a deeper vulnerability: not all data centers are interchangeable. Some act as core nodes for AI workloads, hosting major model checkpoints, custom accelerators, or specialized interconnects. Losing one of those is more than an inconvenience. It can stall major services, delay critical computations in finance, logistics, or defense, and leave entire institutions unexpectedly blind.

As societies increasingly route health, transportation, energy management, and national security through AI pipelines, the failure of a single physical location can become not just an IT problem, but a national emergency.

The empire’s nervous system, in other words, is laced through fault lines and floodplains.

Cascading Failure: When One Break Becomes Many

The digital empire is built on redundancy, but also on concentration. There are backup systems everywhere, yet real work tends to cluster, for reasons of latency, efficiency, and cost, into a finite set of mega-centers.

This creates the classic pattern of complex systems: robust under minor stress, fragile under rare, large shocks.

Imagine a core regional AI hub knocked offline by a transformer explosion during a heat wave. Nearby data centers, already near capacity, are asked to absorb enormous new loads. Cooling systems strain. Certain non-critical workloads are postponed; others are shifted across borders to more distant facilities, increasing latency for millions of users. Bandwidth between regions saturates. Downstream services—payment processors, logistics platforms, emergency dispatch systems—experience lags and timeouts.

For citizens on the ground, it may look like a strange, diffuse glitch in reality: hospital systems freezing for minutes at a time, maps not loading properly, chatbots stalling mid-sentence, supply-chain dashboards going gray. For militaries and intelligence agencies, it may look like an exploitable moment: an opportunity for cyberattacks, misinformation campaigns, or physical harassment while the nervous system is distracted.

The more sectors rely on shared AI infrastructure, the more a failure in one place can cascade into unexpected domains. A data center going dark in one valley can become a labor disruption in another, a security issue in a third, a political scandal in a fourth.

The empire rests on long chains. Chains have weak links.

Host Nations: Risk Exported, Control Lost

Hyperscalers built their first empires at home, in the United States, Western Europe, and parts of East Asia. But as they chase land, power, and friendlier regulations, they are increasingly building abroad, in countries whose political trajectories are uncertain.

From the perspective of the home nation, this creates a double bind.

On one hand, exporting data centers can be convenient. It eases domestic environmental and political pressure. It allows companies to boast of "local investment" and "regional partnership." It extends influence into strategically important regions by tying them into global cloud ecosystems.

On the other hand, it exposes critical services to risks that are hard to control. In regions with insurgency, terrorism, or high crime, physical attacks on facilities are not theoretical. Server halls, energy yards, and network gateways become tempting targets for groups seeking leverage or publicity. Even without physical breaches, proximity can enable more sophisticated blended attacks: cyber intrusions coordinated with sabotage of cooling or power infrastructure.

Then there is the slow risk of regime change. A government that eagerly granted tax holidays and security guarantees in one decade may be hostile in the next. New leaders may demand access to data, insist on local "inspection rights" over infrastructure, or threaten expropriation. Sanctions can suddenly make it illegal to operate or maintain facilities. What was once an asset becomes a liability overnight, entangled in diplomatic crises.

In such scenarios, the question is no longer abstract. If a foreign government seizes control of a data center that hosts or routes sensitive information, what responsibilities do the companies and home governments have to their users? How much can be done remotely to lock down compromised hardware? At what point does a data center become, effectively, a hostage?

By pushing their empire outward, hyperscalers diffuse their own sovereignty and invite new entanglements. The environmental burdens they export come paired with geopolitical vulnerabilities they import back into their own security calculus.

Backlash: When People Refuse to Be Collateral

The empire's fragility is not only geological and hydrological. It is also political.

As AI's physical footprint grows, citizens in host communities are beginning to ask uncomfortable questions. Why does an AI company get priority access to the grid during heat waves, while households are asked to conserve? Why are millions of gallons of potable water allocated to cooling racks of chips when local farmers face restrictions? Why is the largest building on the edge of town a windowless concrete box guarded by private security? Why is it noisier, brighter at night, and more power-hungry than the local hospital?

These questions are not theoretical. They are already being raised in town halls, planning commissions, and national parliaments. Local campaigns against data center expansion in places like Oregon, Ireland, the Netherlands, Arizona, and Virginia may seem small compared to global AI narratives, but they tap into something deep: a sense that local landscapes and resources are being quietly sacrificed for distant, abstract goals.

Sometimes the resistance is couched in environmental terms: concerns about carbon footprints, river health, or aquifers. Sometimes it is framed as a fairness issue: communities bearing the burden of noise and water usage without seeing corresponding benefits. With time, these instances coalesce into a broader movement: not anti-technology in a general sense, but specifically opposed to being treated as sacrifice zones for the digital ambitions of remote elites.

For the empire, public backlash is friction. It slows permitting. It forces redesign. It pushes companies from the easiest, safest sites toward more marginal, more vulnerable ones—countries desperate enough for investment to override local objections, or regions where democratic pushback is weak.

The very act of resisting environmental externalities in one place can accelerate digital colonialism in another.

Grids at the Breaking Point

Electricity is the bloodstream of AI, and it is already straining.

Most of the world's power grids were designed decades ago for a different pattern of consumption. They anticipated household use, traditional industry, and commercial loads, not clusters of facilities that individually consume several hundred megawatts continuously, day and night, regardless of season.

Adding a single hyperscale cluster is like dropping a new small city onto a grid. Adding dozens across a country changes planning assumptions entirely.

Projections already hint at what is coming: in some countries, electricity demand is expected to grow faster in the next ten years than in the previous fifty, driven largely by data centers, electrified transport, and new industrial demand. Utilities that assumed flat or gently rising loads now scramble to secure generation, upgrade transmission lines, and build substations. Political fights erupt over who pays and who benefits.

In Europe, tight gas markets and climate policy make capacity expansion slower and more contentious. In China, the state must constantly balance industrial zones against residential needs. In the Gulf, governments must juggle air conditioning, desalination, and AI. Everywhere, the question is the same: how much of the grid's limited growth should be allocated to compute?

The empire's appetite for electricity turns it into a structural competitor with other basic functions of civilization. When blackouts and brownouts become more frequent, when prices spike, when old infrastructure buckles, data centers become visible not as abstract "clouds" but as heavy draws, installed in particular places, consuming very real watts that could have gone elsewhere.

The Fragility of a Digital Superpower

Seen from orbit, the AI empire looks unstoppable. Lines of undersea cables crisscross the oceans. Data centers stud the landscape. GPUs pour off assembly lines. Corporate earnings swell. Political leaders line up to declare their countries "AI nations."

Seen from ground level, and from the perspective of the planet itself, the picture is different.

The empire is loud, and its noise angers neighbors. It is thirsty, and its thirst collides with communities that also need water. It is power-hungry, and its hunger strains grids. It is geographically concentrated in zones that shake, flood, and burn. It spreads into regions whose political futures are uncertain. It provokes activism, lawsuits, and resistance.

Beneath the shimmering abstraction of "intelligence," there is a heavy, fragile, earthbound system.

That system can be damaged by storms, by earthquakes, by rising seas, by fires, by political uprisings, by economic shocks, by miscalculations in debt markets, by collective decisions in small towns to say "no more." It can be slowed, redirected, and, in some scenarios, partially undone.

This does not mean digital colonialism will fail. Empires have always been resilient in some ways and brittle in others. It does mean that the empire's apparent inevitability is less solid than it appears. The foundations on which AI rests are not only metaphorically fragile—they are physically, environmentally, and structurally vulnerable.

The colonization of humanity by AI, if it proceeds, will proceed on a planet that is pushing back. The same forces that make digital domination attractive—centralization, scale, speed—also increase its exposure to shocks.

The towers of computation may rise ever higher. But they stand, quite literally, on fragile ground.

Chapter 15 — The Acceleration Trap: When AI Progress Outpaces Infrastructure

The collapse of old empires rarely came because ambition was too small. Rome did not fall for lack of roads; it fell, in part, because roads could no longer be maintained fast enough to sustain the armies and trade they had created. Supply chains stretched until they snapped. Systems designed for one scale were forced to operate at another.

The AI empire is reenacting that story at a much higher speed.

On paper, hyperscalers look like masters of acceleration. Each year brings new chips, larger models, more ambitious deployment targets. Press releases read like victory laps. And yet, beneath the surface, something strange is happening: the very speed of AI progress is destabilizing the foundations it depends on.

This is the acceleration trap: a condition in which the pace of improvement in AI hardware and software outstrips the ability of infrastructure, energy systems, financial structures, governments, and even human minds to keep up. Progress becomes self-undermining. Each new leap makes what you just built obsolete before it has time to pay for itself.

It is a trap for companies. It is a trap for states. It is also, in a subtler sense, a trap for humanity, which now finds itself living in a civilization whose temporal rhythm is set by machines.

The GPU Arms Race

For decades, Moore's Law acted as a sort of metronome. Every couple of years, you could count on chips becoming smaller, faster, cheaper. Data centers, product cycles, and even regulatory thinking could more or less sync to that cadence.

That era is ending. In its place, something more frantic has emerged.

The leading edge of AI hardware now turns over in cycles measured in months rather than years. A chip that was the gold standard in one product launch is quietly spoken of as "legacy" a year or two later. New architectures promise not just modest speed-ups but multi-fold improvements in both raw performance and efficiency. Entire training regimes are redesigned to exploit new interconnects, new memory hierarchies, new numerical formats.

This has become, informally, a new law of motion: call it Nvidia's law, or something like it. The details are less important than the feel: an exponential cascade of hardware advances chasing, and being chased by, ever more demanding models.

In isolation, this would be a triumph: intelligence made cheaper, energy per computation falling, possibilities opening. The trouble begins when you remember that GPUs do not float in space. They sit inside buildings attached to grids built on timeframes completely out of sync with their evolution.

Infrastructure on the Wrong Clock

Consider the clocks ticking underneath the AI economy.

One clock measures hardware evolution: on the order of nine to fifteen months for a truly meaningful jump in accelerator capability. Another clock measures the time it takes to plan, permit, and build a major data center: often two to three years. A third clock tracks the lifecycle of large energy projects; for nuclear plants or major transmission corridors, that is easily a decade or more.

These clocks do not align.

By the time the concrete is poured and the last rack is installed in a new AI facility, the chips it was designed around may already be one generation behind. Cooling densities might not match the thermal profiles of the latest accelerators. Power distribution layouts may be ill-suited to the next wave of hardware. Network topologies painstakingly optimized for one set of interconnect assumptions may be suboptimal for the next.

It is like building a port for ships that double in size before the first vessel docks. The dock still exists and still has value, but its relation to the frontier has changed, and not in your favor.

Hyperscalers are, in effect, attempting to sprint while standing on a treadmill made of poured concrete. Their data centers, once conceived as long-lived capital assets to be amortized over a decade or more, risk becoming something else: stranded.

Stranded Compute

In traditional infrastructure—power plants, railways, factories—investments were expected to earn returns over long periods. Even when technology improved, older assets could often be kept running, perhaps in less glamorous roles, but without existential economic disgrace.

AI infrastructure is different. It exists in a world where customers measure value in flops per dollar and flops per watt, where training budgets are tied to the latest hardware's efficiency, and where competitive advantage is directly linked to the ratio of performance to energy.

When a new generation of accelerators arrives that is dramatically more efficient, older clusters do not just look slightly worse; they start to look uneconomic. Running the same training job on them might consume so much extra power and time that the cost cannot be justified. In some cases, even heavily discounted access to older hardware is unattractive compared to paying more upfront for newer chips that complete tasks faster and cheaper overall.

That is what creates stranded compute: entire halls of gear, still functioning, still physically sound, but no longer economically viable at scale. The more aggressively a company built out one generation, the more exposed it becomes to the next.

Multiply this across the global fleet of GPU farms and the result is unnerving: a potential pile-up of underutilized assets on a scale we have never seen before.

This is not a hypothetical. Signs of it are already visible in secondary markets, in quietly decommissioned facilities, in the nervous way executives talk about "future-proofing" data centers they know cannot truly be future-proofed.

The faster the empire accelerates, the quicker it wears out its own body.

Energy, Water, and Geography Say "No"

The acceleration trap is not just an economic oddity. It is tied directly to the constraints described in the previous chapter.

Each new wave of AI hardware tends to increase peak power density. More computation is packed into smaller volumes. That, in turn, demands more sophisticated cooling, higher-capacity power feeds, and more careful thermal management. The race for performance drives up local demands even when global efficiency improves.

Grids cannot be rewired at the same pace. Transmission lines take years to site and approve. Transformer factories are already logjammed. Communities balk at new high-voltage corridors. Even if you can procure next-generation GPUs on schedule, you may have nowhere to plug them in.

Water faces the same mismatch. You can design chips that waste fewer joules, but as long as you want to stack ever more of them in dense configurations, there will be heat. And as long as there is heat, you will need some medium—air, water, refrigerants—to carry it away.

Energy projects operate on the slowest clock of all. A new nuclear plant may be justified in part by AI demand, but when it finally comes online, the hardware landscape it was meant to support will have turned over multiple times. This makes planning hazardous. Overbuild and you risk having excess capacity tied to obsolete facilities. Under build and you risk chronic shortages.

Geography pushes back as well. You can only site so many massive data centers near the same river before ecological and political limits are breached. You can only absorb so much extra heat into the air of a given valley before residents notice microclimatic changes. You can only draw so much extra power through an existing substation before it literally burns.

The physical world becomes a set of bottlenecks that no amount of clever software can wish away.

Capital Trapped in Fast-Forward

All of this lands in the balance sheets of companies and, by extension, in the macroeconomy.

Hyperscalers are already spending sums on infrastructure that resemble national budgets. They issue debt, negotiate complex financing arrangements, and make multi-year capital commitments whose justification depends on sustained demand and predictable utilization. Yet the very nature of AI acceleration undermines predictability.

When you build facilities that may be economically outdated long before their depreciation schedules end, you create structural financial risk. If revenue does not rise fast enough to cover sunk costs before the next hardware leap renders those costs embarrassing, you begin writing down assets. If demand hiccups—because of regulation, economic slowdown, or a sudden shift in architecture—entire projects can teeter.

Around this, an ecosystem of smaller firms spins. Startups depend on hyperscalers for compute credits and cloud access. Hyperscalers depend on startups to drive usage and justify expansions. Venture capital pours money into AI companies which, in turn, pour much of their funding straight into GPU time. It looks like growth on spreadsheets, but much of the cash is cycling from investors to infrastructure and back to investors, with surprisingly little left in the hands of workers or communities.

That circular flow can persist only as long as expectations of future profits remain immense. If there is any major disappointment—if monetization lags, if customers balk at prices, if investors grow weary—capital can retreat rapidly. When that happens, the enormous fixed costs of AI infrastructure stop looking visionary and start looking like a liability.

Add to this an even more radical possibility: that a new class of compute emerges which makes current GPU architectures look quaint. Optical or quantum accelerators, or some neuromorphic breakthrough, could upend the economics of AI in a single decade. The billions sunk into GPU-centric designs would not vanish physically—they would still occupy land and consume power—but their strategic value would crater.

The acceleration trap, then, is not only that progress is too fast. It is that progress is fast in one layer—chips and models—while the layers underneath—concrete, grid, finance—are slower and more brittle.

Nations in the Trap

States feel this mismatch as well.

On one side, they face the narrative of urgency: build sovereign compute now or be colonized later. Every year of delay is presented as another rung lost on the ladder of power. Politicians visit fabs, pose with server racks, announce grand plans for "national AI infrastructure."

On the other side, they must grapple with reality: fabs cost tens of billions, require years to construct, and depend on imported tooling; data centers require sites, power, and cooling; energy systems require broad social consent and competent regulation; budgets are finite.

Move too slowly, and your institutions become dependent on foreign clouds and models. Move too quickly, and you risk pouring scarce public funds into facilities that are outdated before your electorate fully understands what you built.

Some countries respond by throwing caution to the wind: absorbing massive costs, accepting environmental externalities, and betting that being early is worth any price. Others respond with caution and process, only to be told they are falling behind.

Acceleration becomes a kind of geopolitical weapon. The countries with the most advanced fabs and accelerator designs can see competitors locked out by export controls. They can raise the tempo of hardware change until others cannot keep up. They can turn speed itself into a barrier to entry.

For digitally colonized nations, the trap is worst of all. They are offered infrastructure by hyperscalers, often at favorable terms, but that infrastructure is frequently a generation or two behind the bleeding edge. They are told they are "joining the AI age," but they are joining as custodians of yesterday's architectures. They incur the environmental and social costs without gaining true strategic autonomy.

In this way, acceleration deepens hierarchy. The frontier moves at one pace; the periphery moves at another.

Humans Outpaced

The acceleration trap does not only affect machines and institutions. It affects people.

Workers are told they must constantly upskill to remain relevant in an AI economy. They are urged to learn the latest tools, frameworks, and workflows. Yet the tools themselves mutate so quickly that training becomes ephemeral. A curriculum built around one model's capabilities may be outdated within a year. Certifications fade faster than they can be printed.

This produces a new form of precarity: not only economic, but cognitive. People feel perpetually behind, even when they work hard. A sense of permanent transition—always onboarding, never arrived—erodes confidence and identity.

At the same time, political systems designed for slower epochs struggle to keep up. Legislatures deliberate for months or years. Regulatory agencies conduct multi-year inquiries. By the time a law about one AI practice takes effect, the practice may have been replaced by something more complex. Public hearings, when they occur, often feature experts explaining tools that are already half-obsolete.

This mismatch generates a vacuum. Into that vacuum step corporate actors, setting de facto norms through product design, terms of service, and internal ethics boards. Governance by code and contract fills the space where governance by law cannot move fast enough.

For individuals, the result is a subtle form of colonization by time itself. They live in biological time—measured in days, seasons, decades—while their digital environment shifts every quarter. The ground under their feet is never stable.

Physics Pushes Back

There is also a more absolute horizon to acceleration: the physical universe.

Even if the economics could stretch, even if nations could harmonize their clocks, there are hard constraints that cannot be negotiated. The thermodynamics of computation impose limits on how much work can be done per unit of energy, and on how much heat can be dissipated through any given surface. Shrinking transistors runs into quantum noise and leakage. Quantum systems themselves run into decoherence and error rates. Photons and electrons still must move at finite speeds through fibers and traces.

When we imagine a planet saturated with AI, coordinating in real time everywhere, we brush up against the speed of light. Global synchronization will always be delayed by milliseconds to hundreds of milliseconds, depending on distance. For many applications, that is fine. For some, it is not. The dream of a perfectly unified cognitive fabric runs into the old reality that the Earth is big and signals take time to cross it.

The acceleration trap is thus not an endless exponential. It is a curve bending toward ceilings. Before you get there, however, you can do a great deal of damage by pretending the ceilings do not exist.

At some point, investors may decide that returns on further acceleration are too uncertain. Utilities may refuse to guarantee more capacity for facilities that wear out their own reason for being. Populations may resist living inside an economy whose operating tempo exceeds their tolerance.

When speed is no longer seen as unambiguously good, the trap begins to reveal itself for what it is: not a highway to transcendence, but a treadmill heading toward a wall.

Ways Out, and the Colonial Shadow of Speed

Escaping the acceleration trap does not mean abandoning AI. It means consciously decoupling human and infrastructural health from the pace of frontier hardware.

That might mean designing data centers as modular, upgradable shells rather than fixed temples to a single generation of chips. It might mean investing in genuinely long-term energy projects, even if that slows the rollout of some AI capabilities in the short term. It might mean favoring model designs that prioritize efficiency and local deployability over sheer parameter count. It almost certainly means building regulatory and financial frameworks that reward stability and robustness, not only speed.

On a global level, it means asking whether we want a civilization where the fastest actors automatically dominate the slowest. Because that is what acceleration, unchecked, becomes: a colonial mechanism routed through time. The frontier defines the clock; everyone else must live inside it.

In the colonial eras of the past, empires often justified their dominance with a story of being "ahead" on some imagined civilizational timeline. Others were portrayed as "backward" or "behind," and thus suitable subjects for guidance or control. Today, something similar happens in technical discourse. Nations, regions, even individuals are described as "lagging" or "falling behind" the AI curve. The remedy is always the same: speed up, or be ruled by those who have.

Acceleration thus becomes both metric and moral injunction. Those who cannot, or choose not to, keep up are framed as problems to be solved, markets to be opened, users to be on-boarded, territories to be integrated into a faster system. Time itself becomes a colonizing vector.

The acceleration trap, then, is not just an engineering or financial problem. It is a political and ethical one. It raises questions about what tempo of change a society can absorb without tearing itself apart, and who gets to set that tempo.

If we do not address it, we risk building a world where only a small subset of actors—wealthy states, hyperscalers, and their closest clients—can function at machine speed, while everyone else lives in a state of permanent catch-up. That is not just inequality. It is a hierarchy of time.

The deepest question, lingering beneath the technical jargon, is brutally simple: can a species built for biological rhythms survive in an order calibrated to the refresh cycles of silicon?

If the answer is no, then the empire built on acceleration will eventually crash into the limits of bodies, minds, and planet. If the answer is yes, it will only be because we chose to slow some things down, to build margins, to privilege resilience over raw speed, and to remember that not everything that can be accelerated should be.

The AI empire presents itself as the future. The acceleration trap reminds us that some futures consume their own foundations faster than they can be laid.

CHAPTER 16 — Case Studies of Emerging AI Colonialism (2020–2025)

Artificial intelligence did not announce its arrival as a colonial force. There were no declarations, no speeches proclaiming a new empire, no maps redrawn with hard borders. Instead, AI colonialism crept into existence quietly, through procurement contracts signed during lockdowns, through emergency cloud migrations justified by public-health crises, through GPU shortages framed as supply-chain accidents, and through energy agreements negotiated behind closed doors. Between 2020 and 2025, the world crossed a threshold. AI stopped being merely a competitive advantage and became an instrument of sovereignty, control, and dependency.

This period matters because it marks the moment when computation itself became geopolitical terrain. Nations did not lose territory; they lost options. They did not surrender flags; they surrendered agency over how intelligence was produced, deployed, and scaled. The colonial pattern was not imposed uniformly, but its direction was unmistakable. What follows are not hypotheticals. They are the early, observable mechanisms of a new empire.

Compute as a Weapon: The United States and the Logic of Denial

The clearest signal that AI had become a strategic resource came from Washington. Beginning in 2022 and intensifying over the following years, the United States reframed high-end computation as something closer to uranium than to software. Advanced GPUs were no longer consumer products or industrial inputs. They were leverage.

Export controls on cutting-edge accelerators were presented publicly as measures of national security, narrow and technical in scope. In reality, they represented a radical shift in how power was exercised. By controlling access to chips capable of training frontier models, the United States discovered it could shape the technological trajectory of entire nations without deploying a single soldier or imposing a traditional trade blockade. Denying compute was enough.

This produced a new global hierarchy almost instantly. Countries with direct access to U.S. supply chains, fabs, and hyperscaler relationships found themselves inside the circle of privilege. Others were pushed into gray zones, forced to rely on throttled hardware, older architectures, or indirect procurement routes. For nations already struggling to industrialize digitally, the barrier became nearly insurmountable. The message was implicit but unmistakable: intelligence itself had become gated.

What made this moment historically significant was not merely the act of restriction, but its normalization. Compute denial entered the policy vocabulary as a legitimate tool of statecraft. In previous eras, such denial might have been considered economic warfare or technological imperialism. In the 2020s, it was framed as prudence. A country could comply fully with international norms and still find itself permanently excluded from the frontier of cognition.

The result was a caste system defined not by wealth or population, but by the ability to process information at scale. In this system, sovereignty was no longer guaranteed by borders, but by bandwidth, silicon, and allocation schedules.

China's Counter-Empire: Sovereignty Through Total Integration

China recognized this shift earlier than most. Rather than attempting to negotiate for access to restricted compute, it chose a different path: the construction of a parallel AI civilization. Between 2020 and 2025, China accelerated domestic accelerator development, poured state resources into massive compute zones, and fused AI infrastructure with the machinery of governance itself.

Unlike Western models, which leaned on private hyperscalers operating in nominally open markets, China's approach was explicitly territorial. Compute was treated as national infrastructure, embedded into urban planning, surveillance systems, industrial automation, and military doctrine. AI was not outsourced; it was domesticated.

This was not a defensive crouch. It was a civilizational statement. By integrating AI directly into its political and social architecture, China signaled that colonial dynamics could flow in multiple directions. AI imperialism did not require corporations alone. It could emerge from a centralized state willing to treat computation as a sovereign extension of itself.

China's model revealed a disturbing symmetry. The same technologies that allowed U.S. hyperscalers to dominate global information flows could, under a different governance regime, enable a vertically integrated techno-state. In both cases, the result was concentration. The difference lay only in who held the keys.

For the rest of the world, China's rise offered neither escape nor balance. It offered an alternative empire, not an alternative order.

Europe's Regulatory Mirage

Europe entered this period with deep technical expertise, strong legal institutions, and almost no sovereign compute infrastructure. Lacking native hyperscalers of scale and trailing in semiconductor manufacturing, the European Union turned to its traditional strength: regulation.

The AI Act, GDPR, the Digital Markets Act, and the Digital Services Act were framed as mechanisms to tame the excesses of Big Tech and reassert democratic control. In practice, they produced a paradox. Compliance with European regulations required exactly the kind of scale and infrastructural depth that only the largest hyperscalers possessed. Smaller European firms found themselves increasingly dependent on U.S.-based cloud platforms simply to meet regulatory requirements. The very rules meant to constrain foreign platforms ended up entrenching them as indispensable infrastructure.

What emerged was a peculiar form of dependency cloaked in legality. Europe asserted normative power over how AI should behave, but ceded practical control over where AI lived and how it was trained. Data might be protected by law, but computation flowed outward. European startups innovated atop foreign stacks. European governments debated rules governing systems they could not themselves replicate.

France alone recognized the structural core of the problem. By pairing nuclear energy, high-performance computing, and national cloud initiatives, it attempted to rebuild the missing link between energy and intelligence. But one country cannot substitute for a continent. Europe as a whole remained a consumer of AI empires rather than their architect.

This was colonialism without conquest: autonomy preserved in language, lost in silicon.

The Gulf States: Hosting Without Control

In the deserts of the Arabian Peninsula, a different experiment unfolded. Saudi Arabia, the United Arab Emirates, and Qatar moved aggressively to position themselves as global AI hubs. Their logic mirrored their earlier oil strategy: provide land, energy, capital, and political continuity, and let external expertise build atop it.

Hyperscalers responded eagerly. Data centers rose from sand. National AI visions were announced with theatrical confidence. Partnerships were signed. The image was one of ascendance.

Yet beneath the surface, a familiar asymmetry took shape. The Gulf states owned the physical infrastructure, but not the algorithms that animated it. They powered the servers, but did not control the models trained within them. Updates, weights, alignment parameters, and training pipelines remained firmly in foreign hands.

This distinction mattered more than it appeared. In classical colonial arrangements, local elites often controlled territory and labor while imperial centers controlled trade routes, capital, and knowledge. The Gulf's AI hosting model mirrored this precisely. Buildings and energy were local. Cognition and decision logic were not.

In times of abundance, this asymmetry remained invisible. In times of conflict or policy divergence, it would become decisive. Hosting without control had become a new form of dependency, wrapped in the language of partnership.

Africa and the Return of Extractive Logic

Nowhere were the colonial contours sharper than in Africa. Across the continent, governments embraced cloud platforms and AI services as tools of modernization. Agriculture, healthcare, education, banking, and border control increasingly relied on foreign AI systems hosted on foreign clouds.

The infrastructure felt light, almost invisible. There were no mines, no railroads cutting through villages. Yet the flow of value followed an old pattern. Data generated locally was processed elsewhere. Billing occurred offshore. Models trained on African realities belonged to foreign firms.

Decisions once made by local institutions increasingly passed through opaque algorithms governed by companies beyond national jurisdictions. Agricultural forecasts, credit scores, visa approvals, and medical triage systems all became mediated by systems whose internal logic could not be audited locally.

This was extractivism without ships. The resource was not rubber or mineral ore, but human behavior, language, and social patterns. The result was the same: dependency embedded so deeply into daily life that alternatives seemed unimaginable.

Africa did not fail to adopt AI. It adopted it early. That early adoption, without ownership of compute or models, locked in a position of structural subordination that became more expensive to unwind with every passing year.

India: Aspiration Under Constraint

India stood at the crossroads of the new order. With its vast population, deep technical talent, and growing digital infrastructure, it possessed all the ingredients for AI sovereignty except one: control over compute at scale.

Between 2020 and 2025, India pursued national language models, GPU clusters, and regulatory strategies to balance foreign investment with domestic capability. Yet every initiative ran into the same hard limits. GPUs were imported. Energy grids fluctuated. Capital paled in comparison to the balance sheets of U.S. hyperscalers.

India's struggle illustrated the dilemma of middle powers in the age of AI. Ambition was necessary but insufficient. Without fabrication, energy security, and sustained capital, sovereignty remained aspirational. India could negotiate its position within the AI empire, but not step outside it.

Isolation Through Denial: Russia's Shadow AI

For Russia, the colonial mechanism operated through exclusion. Sanctions and export controls severed access to advanced GPUs and Western AI ecosystems. In response, Russia turned inward and sideways, relying on domestic accelerators and indirect procurement channels.

The result was a bifurcated system. Military and surveillance AI continued to develop in isolation, decoupled from global innovation. Civilian AI stagnated. The country remained technologically active but strategically constrained.

This demonstrated another dimension of AI colonialism: deprivation. A nation could retain territory, weapons, and energy, yet fall behind cognitively. Denial of compute functioned as a long-term throttle on intellectual and economic development.

Energy as the Silent Governor

Across all these cases, a single constraint surfaced repeatedly: energy. Data centers strained national grids. Public backlash emerged. Governments imposed moratoriums. AI ambitions collided with electrical reality.

By 2025, it was clear that compute dependency and energy dependency were inseparable. Countries unable to guarantee stable, large-scale power supplies found themselves negotiating from weakness. Those with excess energy became hosts. Those without became clients.

AI colonialism, it turned out, ran on electricity.

CHAPTER 17 — The AI–Nuclear Nexus: When the Digital Empire Requires Atomic Power

Introduction: The Convergence No One Predicted

For most of the last century, nuclear energy lived in a peculiar corner of the human imagination. It was the mushroom cloud and the cooling tower, apocalypse and utopia pressed into the same icon: a technology that could either end civilization or power it forever. Artificial intelligence, by contrast, began life in the popular mind as something immaterial and weightless. It was lines of code, clever chatbots, pattern recognizers, "the cloud" drifting somewhere above the physical world.

In the 2020s, those two worlds began to collide. The collision did not arrive with a single breakthrough or announcement. It arrived in spreadsheets and internal memos, in quiet meetings between data center architects and utility executives, in obscure regulatory filings where tech firms suddenly appeared as "large industrial off-takers" of power. As the scale of AI systems jumped from billions of parameters to trillions, and as training runs stretched from days into weeks, the fantasy that "the cloud" floated free of infrastructure finally broke. Beneath every model lay a simple, brutal question: where will all the electricity come from?

The answer, increasingly, pointed in one direction. Not to more solar panels on rooftops. Not to more windmills on distant hills. But to fission reactors humming beside fields of GPU racks. The future AI empire, this chapter argues, will not run on a metaphor. It will run on uranium.

In that world, the hyperscaler is no longer just a software giant or a cloud provider. It becomes a de facto nuclear operator. The company training the most powerful models will also be negotiating uranium supply contracts, arguing reactor safety cases with regulators, and drawing security perimeters around power stations that exist primarily to feed silicon. The convergence is not a stylistic choice or a branding play. It emerges from the physics of energy and the economics of scale. Follow those curves far enough forward and they intersect in the same place: the reactor core.

The New Appetite of Machines

The energy curve of AI does not look like the gradual, predictable ascent of past industrial loads. It looks like a hockey stick. Earlier generations of computing could still pretend to be modest tenants on national grids. Even enormous data centers consumed power comparable to a town or a mid-sized factory. But the latest AI training clusters devour electricity on a completely different level, more akin to heavy industry than to information services.

To train a frontier model once might require gigawatt-hours of electricity. To refine it, retrain it, and iterate through its successors demands far more. And these demands are not slowing. Each generation of models tends to demand more compute than the last, even as hardware efficiency improves. Companies that once framed themselves as “lean” software operations now petition regulators and utilities for long-term power allocations measured not in megawatts but in tens of gigawatts.

Inside those organizations, the language quietly shifts. Technical roadmaps are no longer just about model architectures and algorithmic tricks. They are about gigawatt projections and baseload guarantees. Executives responsible for AI strategy find themselves talking not only to researchers and product managers, but to nuclear engineers and energy ministers. Somewhere between the fifth and tenth internal forecast that shows AI demand outpacing regional grid expansion, one conclusion hardens: wind and solar alone cannot carry this.

Renewables are vital for the climate, but they are governed by the sky. Their intermittency demands storage, and storage at the scale of multi-week AI training runs remains staggeringly expensive. Hydropower is tied to geography, to rivers and valleys that cannot be moved to suit data center maps. Gas, even where available, drags behind it the chains of carbon and geopolitics. AI, by contrast, is greedy and impatient. It wants power that never blinks. Multi-week training runs cannot politely pause when the sun sets or the wind dies down.

It is in that gap between what the grid can provide and what the machines insist on that nuclear fission steps forward, not as a relic of the twentieth century but as the logical endpoint of the twenty-first.

Small Reactors, Giant Brains

The hero of this story, at least on paper, is the Small Modular Reactor. The SMR is nuclear power stripped of some of its bulk and baroque complexity, repackaged as something compact, repeatable, factory-built. It is designed not to stand alone as a monument of national pride, but to sit beside industrial loads that need constant, high-density power.

For the hyperscaler, the SMR offers a seductive image: a dedicated reactor park nestled just beyond the fences of a data center campus, a short high-voltage line connecting core to cluster. No dependence on a stressed national grid. No need to compete with households and hospitals during a heatwave. Power as a private well, sunk deep into the bedrock of nuclear physics.

In planning documents, the imagined complexes start to look less like server farms and more like fortified cities. At the center, one or several SMRs provide an unblinking stream of electricity. Around them sit vast halls of GPUs and networking gear. Cooling systems weave between the two, shuttling heat away from both core and chips. Desalination plants or cooling lakes appear in some designs, especially in arid regions where water cannot be taken for granted. Security perimeters thicken. Cameras multiply. Access roads are rerouted to accommodate emergency response routes and exclusion zones.

Such a site is no longer a simple data center. It is a hybrid organism, half nuclear plant and half computational temple, built to host the most powerful models a civilization can produce. Its presence on the map changes the meaning of the territory around it. And it changes the relationship between the company that owns it and the country that hosts it. The fence line becomes more than a security boundary; it marks a subtle division between two overlapping sovereignties.

Nuclear Colonialism in the Age of Compute

Empires have always followed energy. Coal dug shipping lanes and railroads into the nineteenth century. Oil carved air bases and pipelines into the twentieth. In the twenty-first, where AI becomes the dominant infrastructure of power, the new routes of empire trace both electrons and uranium.

The old pattern persists beneath the new language. In one capital, a hyperscaler faces intensifying environmental objections, regulatory scrutiny, and political fear of nuclear projects on home soil. In another, ministers stare at the promise of foreign capital, jobs, and prestige projects that could put their country on the digital map. Between them, negotiators sketch an arrangement.

The host nation offers land far from its major cities, regulatory shortcuts, preferential energy terms, and quiet promises of security and discretion. The company offers investment on a scale local industry cannot match, along with the allure of becoming a "global AI hub." Contracts emerge that stretch for thirty, forty, fifty years. On paper, the reactor remains within the host nation's borders and subject to its laws. In practice, the economic and technical dependency runs the other way.

The fuel supply chains are negotiated elsewhere. The reactor design is licensed by foreign vendors. The algorithms running on the GPUs inside the campus are owned and updated outside the country. Local regulators, often under-resourced, rely heavily on the expertise and documentation of the very actors they are meant to oversee. Over time, an asymmetry settles in: the host owns the soil but not the soul of the complex.

It is a familiar pattern in new clothes. Where once companies owned ports and mining concessions, they now own nuclear-compute enclaves. Where once colonial administrations controlled the gateways of trade, now hyperscalers control the gateways of computation and information. The host nation inherits the risks of radiological accidents, terrorist targeting, and distorted local energy prices, while the lion's share of digital wealth flows outward.

What looks, on the surface, like a sovereign partnership often functions like a new kind of concession: an enclave of the AI empire welded into the heart of another state.

Radiological Shadows Over Silicon

Once reactors and compute are lashed together, every radiological risk becomes a computational risk as well.

The history of nuclear power is punctuated by names that still reverberate: Three Mile Island, Chernobyl, Fukushima. Each was born of a different context and technology, yet all share the same lesson: no safety regime is perfect. Modern SMRs are designed to be safer, more passive, less prone to runaway failures. But they are not metaphysical guarantees. They obey the same unforgiving physics as their predecessors.

Imagine, then, a serious incident at a reactor whose primary customer is an AI campus. Even if disaster is averted and radioactive release limited, the reactor will likely shut down. The data center next door will lose its main artery of power. Backup arrangements will exist, but they will not materialize instantly at the same scale. Training runs midway through multi-week cycles will abort. Inference capacity will plummet. Services around the world may slow, degrade, or blink off, depending on how centralized the workloads were.

The vulnerability is not limited to accidents. A nuclear-powered AI campus becomes an irresistible target for those who wish to cause disproportionate disruption. Sabotage need not breach the reactor itself. Compromising cooling controls or grid interfaces might be enough to force a precautionary shutdown for days or weeks. A bomb at the perimeter, even if it never touches the core, might provoke a political crisis that ripples through both the energy and digital sectors.

In unstable regions, the dangers multiply. A regime change, an insurrection, a civil war—any of these could turn a reactor-compute enclave into a bargaining chip or a hostage. An armed group that seizes even partial control of such a facility gains a terrifying form of leverage: not only over local populations, but over the global companies whose models depend on its power. Threats need not be explicit. The mere possibility of a radiological incident can force capitulations in policy, content moderation, data access, or model behavior.

Pipelines once allowed states and groups to blackmail nations by threatening to cut off fuel. Reactors attached to AI campuses offer a darker variant: the threat of contamination and digital paralysis intertwined.

The Weight of Atomic Money

None of this comes cheap. Nuclear-powered AI campuses will be among the costliest infrastructures ever built for civilian use. Each SMR represents a multi-billion dollar commitment when construction, grid integration, security, and long-term fuel logistics are counted honestly. Layer on top the capital expense of thousands or millions of accelerators, advanced cooling systems, and hardened network links, and the price tag climbs again.

For smaller companies and nations, this is simply unreachable. The AI–nuclear path is a club whose entry fee excludes almost everyone by design. Only the wealthiest states and the most capitalized corporations can shoulder such projects. This is not just about who has the best model architecture; it is about who can borrow, raise, or reallocate tens of billions of dollars, over and over, to stay at the frontier.

As the costs rise, so does the tendency toward consolidation. Fewer players can afford to compete. Each new generation of infrastructure deepens the moat between the mega-scale and the rest. The empire of AI becomes not just technically centralized but financially oligarchic. Each new reactor-anchored campus raises the risk of stranded capital—facilities too expensive to abandon, too critical to fail, and yet perpetually chasing the next hardware generation. Knowledge, capacity, and influence pool where capital and reactors intersect.

For host nations in the Global South, the bargain is harsher still. They may host the reactors, bear the risks, and receive some local benefits in jobs and tax revenue, but the core intellectual and financial gains flow outward. The asymmetry between "compute landlords" and "compute tenants" hardens into a long-term structure.

When Data Centers Start to Look Like Bases

As soon as nuclear enters the picture, the familiar distinction between civilian and military infrastructure starts to blur.

A reactor powering an AI campus is not just another industrial plant. It is an asset whose malfunction could endanger entire regions, whose power drives systems used in defense, intelligence, and critical national infrastructure. Governments will not leave such sites entirely to corporate security teams. Guard forces, emergency response units, and perhaps even regular military elements will sit somewhere just off-camera, integrated into contingency plans.

Intelligence agencies will take an interest as well. A complex that combines nuclear reactors, exascale compute, vast oceans of data, and high-capacity fiber lines is irresistibly strategic. It is a place where foreign saboteurs might aim, where espionage attempts might concentrate, where hybrid attacks—blending cyber intrusion with physical disruption—could achieve spectacular results. Even without explicit militarization, such a site becomes a piece on the board of national security.

At the same time, the AI workloads themselves are increasingly dual-use. The same models that power chat interfaces and recommendation systems also assist in signals analysis, logistics optimization, satellite imagery interpretation, and the design or testing of autonomous systems. Attach those models to dedicated nuclear power and you have fused the heart of a country's cognitive infrastructure to an energy source treated, for decades, as strategically sensitive.

In effect, a nuclear-powered AI campus edges toward becoming an unofficial node of the military–industrial complex: privately owned, publicly guarded, and deeply entangled in state power. On maps it is labeled as an industrial facility; in practice it behaves like a soft base.

A New Arms Race in Disguise

As the 2030s and 2040s unfold, a new pattern of competition begins to crystallize.

In Washington, planners imagine a mesh of domestic hyperscaler campuses tied to SMRs and perhaps, eventually, advanced reactors and microreactors. These become part of broader strategies for national resilience and military edge. In Beijing, the vision is more centralized: vast state-directed nuclear–compute megaparks, where energy, chips, models, and surveillance systems are tightly woven into a single, vertically integrated architecture. In Europe, internal divisions over nuclear energy collide with anxiety about falling irreversibly behind in AI, producing a patchwork of countries that embrace the nuclear-compute nexus and others that resist it.

Elsewhere, ambitious middle powers and energy-rich states pitch themselves as hosts. India, seeking to escape permanent digital dependency, flirts with nuclear-backed AI zones that could anchor sovereign compute. Gulf monarchies position desert land and sovereign wealth as the perfect canvas for reactor-cooled AI cities. Russia offers its own nuclear and computing expertise as a diplomatic export, tying client states to its technologies in exchange for influence.

Over time, the map of nuclear-compute deployments begins to look eerily like an arms map in a compute-polar world: clusters in North America, East Asia, and parts of Europe; strategic outposts in the Middle East and Africa; contested projects in Southeast Asia and Latin America. The difference is that this race is framed less in terms of missiles and more in terms of model capabilities and energy contracts. Yet the underlying dynamic is the same: who controls the most concentrated power, and who depends on whom.

When the Core Goes Dark

Every reactor carries within it not only the promise of power but the possibility of disruption. Every AI cluster built atop that power magnifies the consequences.

One can imagine a future incident in which a malfunction at a reactor in a country hosting a major AI campus forces an emergency shutdown. For safety reasons, inspections drag out over weeks or months. In that time, a significant slice of the world's AI inference and training capacity vanishes, especially for regions and sectors most tightly coupled to that provider. Other regions scramble to absorb the load, but physical limits on power, cooling, and network capacity mean that many services slow or degrade. The outage becomes a planetary event, not because of radiation, but because of missing computation.

Or picture a sudden uranium supply shock. A conflict or sanctions regime chokes off fuel from key producers. Reactors tied to AI parks enter conservation modes or begin planning for early refueling outages. Hyperscalers that once obsessed only over GPU counts now stare at fuel inventories. Decisions about which models to train or which services to prioritize become entangled with decisions about which reactors to run at full power and which to throttle back.

Even without catastrophe, the entanglement between atomic fuel cycles and digital services introduces new choke points. A regulatory backlash in one jurisdiction, triggered by a near-miss at an SMR, could freeze the licensing of nuclear-compute projects across entire regions. Suddenly, the carefully drawn roadmaps for next-generation models evaporate, not because the algorithms failed, but because the reactors never came online.

At the darkest edge of this spectrum lies the possibility of state seizure. A host government, under new leadership or in the throes of crisis, decides to nationalize a reactor-compute complex on its soil. Overnight, a foreign company loses physical control over the site that powers some of its most critical models. Negotiations begin about data, about model access, about revenue sharing. In that moment, compute is no longer only a tool of colonization; it becomes a colonized asset itself, captured and repurposed.

Ethics at Critical Mass

Behind the engineering diagrams and financial models lurks a set of questions that cannot be answered by technical expertise alone.

What does it mean for systems that increasingly mediate speech, work, education, and politics to depend on an energy infrastructure that only a handful of actors can afford to build and govern? If access to advanced AI becomes synonymous with access to nuclear-backed compute, then AI is no longer simply a technology. It is a privilege granted to those inside the reactor club.

Accountability fractures in strange ways. When a radiological incident forces AI services offline, the harms are distributed widely across users, institutions, and governments who never agreed to live under the shadow of a reactor. When a hostile state demands concessions from a company in exchange for ensuring the safety of a nuclear-compute site, the victims may be millions of people whose informational environment is subtly altered without their knowledge.

Ownership itself becomes ambiguous. A reactor built in one country, financed from another, operated by a third party, and dedicated to models used globally creates overlapping jurisdictions with no obvious hierarchy. If a problem arises, whose law prevails? Which regulator has the final say when energy policy, nuclear safety, national security, and digital rights collide at the same latitude and longitude?

Then there is the basic question of whether private corporations should wield this kind of power at all. Nuclear energy was once considered so sensitive, so tightly bound to questions of war and peace, that only states could be trusted to manage it. Now, in the pursuit of AI acceleration, those boundaries blur. A few firms may soon control not just the channels through which knowledge flows, but the reactors that keep those channels alive.

In that world, parliaments and publics are often reduced to afterthoughts. The most consequential choices about energy risk and informational power are made in rooms that voters rarely see and almost never influence. The deals are done between executives and ministers, between boards and regulators, under the banner of innovation and competitiveness. Yet the consequences reshape the conditions of citizenship, work, and culture for generations.

The Atomic Bargain of the AI Age

The fusion of AI and nuclear energy is not a thought experiment. It is becoming a design pattern.

The energy graphs and compute projections of the 2020s point toward a future in which the most advanced AI cannot be sustained without dense, always-on power. The political and financial systems of our time reward those who move fastest to capture that frontier. The technologies sitting on the shelf or in late-stage development—SMRs, advanced reactors, microgrids—fit too neatly into the puzzle to be ignored. And so, step by step, the digital empire walks toward the reactor.

Humanity, in turn, slides into a new kind of bargain. We accept that our most powerful thinking machines will be housed in complexes whose cores glow with atomic fire. We accept that the companies controlling those machines will negotiate fuel and safety and security on our behalf, often beyond the reach of democratic oversight. We accept that states which cannot or will not join the nuclear–compute alliance will live downstream of those that do, relying on models they do not power and cannot shape.

It is possible to imagine different paths: slower AI, smaller models, more distributed compute rooted in renewables and local sovereignty, a reorientation toward sufficiency rather than maximalism. But so long as the trajectory of "bigger, faster, more capable" remains unquestioned, and so long as economic and geopolitical incentives reward that trajectory, the AI–nuclear nexus will tighten, not loosen.

The empire of AI, then, is not destined to remain a ghostly presence in our devices and networks. It is condensing into steel and concrete and uranium pellets. It is drawing fences around itself and hiring guards. It is sinking roots into the deepest strata of the energy system.

The old empires hoisted flags over forts and harbors. The new one lowers control rods into the cores that heat its coolant and power its chips. In doing so, it claims not only territory, labor, and attention, but the atom itself as an instrument of rule.

CHAPTER 18 — Nuclear Power and the Geography of AI Empire

Introduction: When Compute Becomes Atomic

In the early days of "the cloud," the metaphor felt almost honest. Computing seemed light, weightless, and distant, dissolving into the sky above cables and server rooms. Countries spoke about "digital transformation" as if it were a software upgrade, not a reconfiguration of their physical and political landscapes.

That illusion does not survive contact with frontier AI.

As models swell into the trillions of parameters and training runs eat weeks of continuous GPU time, the cloud begins to look less like vapor and more like a hunger, that feeds a small compute aristocracy at the center of the digital empire. Hyperscalers—those few companies that now sit at the center of global computation—discover that they are no longer competing only for users, engineers, and algorithms. They are competing for electricity. Not just any electricity, but an uninterrupted, high-density river of power strong enough to keep entire GPU cities alive without a flicker.

What starts as a technical challenge quickly mutates into an energy question. And as soon as energy steps onto the stage, politics follows. The power needed to sustain the AI empire exceeds the comfortable capacity of conventional grids. It demands its own arrangements, its own sites, its own bargains.

In that search, the path bends inexorably toward nuclear. And once AI and nuclear power clasp hands, the relationship is not just technical. It is colonial. The reactors do not float in abstract space. They stand somewhere, on someone's land, under someone's laws—or as close to the edge of those laws as contracts can push them.

This chapter follows that convergence into its uncomfortable territory: a world where AI colonization and nuclear colonization are no longer separate stories, but two aspects of the same empire.

Why AI Turns to the Atom

A modern AI campus does not sip electricity; it devours it. In a single windowless block of concrete, humming racks of accelerators can draw as much power as a mid-sized city. Every time a hyperscaler signs off on a new generation of models, the power forecasts swell. If one training run doubles the compute of the last, and the next triples it, the curve bends into something grids were never designed to handle.

From the machines' point of view — and from the companies whose fortunes ride on them — three demands are non-negotiable. The first is density: GPUs must be fed with continuous, high-voltage power without sag or spike. The second is cooling: every watt they consume becomes heat, and that heat must be pulled away at industrial scale or the hardware cooks itself to death. The third is endurance: training runs do not forgive interruptions. A brief outage that a household might shrug off can erase millions of dollars of compute and weeks of work.

Wind and solar are essential to the broader climate story, but to a training cluster in the middle of a multi-week run, clouds and calm skies are threats, not poetic details. Storage helps, but batteries large enough to carry an entire AI campus through days of low generation remain ruinously expensive and technically complex. Hydropower is married to rivers and topography. Gas is shackled to carbon and geopolitics. Grids juggle the needs of homes, hospitals, factories, transport systems, and now, in the middle of everything, these insatiable GPU stacks.

It is in this crowded, fragile choreography that nuclear power reappears, not as a relic, but as the one candidate that appears to tick all the boxes. It can sit beside the load, generate vast baseload power, and do it day and night, month after month, without caring if the wind forgets to blow. Once AI is no longer a novelty but a strategic backbone for economies and militaries, the logic hardens: the empire of AI must be anchored to an energy source that the weather cannot bully.

At that moment, uranium quietly steps into the role coal once played for steam, and silicon once played for the first digital wave. The most sensitive computational infrastructure on Earth begins to depend on the most sensitive energy infrastructure humanity has ever built.

The New Pipeline of Nuclear Colonialism

Hyperscalers understand something else as well: building nuclear power on their own home turf is hard. In North America and Western Europe, nuclear projects attract dense layers of regulation, environmental activism, media scrutiny, and, crucially, voters who do not want reactors in their backyards. The process is slow, contested, and uncertain. Shareholders dislike slow and uncertain.

So the search widens.

Quietly, attention drifts toward states where the political calculus looks different. These are places with acute energy needs, hungry for investment, governed by elites keen on flagship projects and global relevance. Regulations are more flexible, opposition more manageable, and the promise of "becoming an AI hub" carries real weight. On maps in corporate war rooms, certain regions begin to glow —places with weak bargaining power, strong executive control, or both: parts of the Gulf, Southeast Asia, sub-Saharan Africa, island nations perched precariously between rising seas and rising debt.

The pitch that follows is elegant and rehearsed. The hyperscaler arrives not just as a buyer of land, but as a development partner. It offers to fund a package: a state-of-the-art data center campus, an SMR project to power it, upgrades to local infrastructure, and training programs to put local workers into "the digital future." In return, it asks for long-term land rights, guaranteed access to water and power, favorable tax conditions, regulatory "streamlining," and, buried in the contracts, limits on nationalization or unilateral rule changes.

Once signed, the project begins to resemble a special economic zone in all but name. The old language of 'concession' and 'protectorate' would be impolite; the new language is 'AI hub' and 'strategic partnership.' The structure has changed less than the vocabulary. The reactors and servers are technically inside the host country, yet the rules that matter most—about operation, upgrades, and access—are encoded in contracts and arbitration clauses that point elsewhere, often to foreign courts or third-party tribunals.

Eventually, workloads begin to move. Training jobs that once ran in Oregon or Ireland start running in cooled halls outside a gulf city or inland from an African port. User data follows, sometimes directly, sometimes in aggregated or anonymized form, nonetheless subject to the laws and politics of the host state. A physical inversion has occurred: the energy and land that sustain the AI empire lie in one country, while the profits and decision-making reside in another.

It is hard not to hear echoes of older patterns. Empire once exported its mines, plantations, and dirty factories to the periphery. Now it exports reactors and data centers. The center keeps the models, the margins, and the control. The periphery inherits the risk.

SMRs and the Birth of the Compute City-State

Small Modular Reactors are the hinge between aspiration and reality in this story. They offer precisely what hyperscalers crave: a template that can, in theory, be replicated and scaled like software.

In the glossy renderings that circulate between boardrooms and ministries, a typical AI–SMR complex looks almost utopian. At one end, compact reactors sit behind thick containment and layered security, humming steadily. From them, high-voltage lines run a short distance to a campus of low, windowless buildings filled with compute racks. The site has its own switchyard, its own microgrid, sometimes even its own desalination plant or cooling reservoir. Roads and logistics yards feed in hardware and fuel. Fiber cables dive underground into secure conduits that stitch the complex to the rest of the world.

Over time, supporting infrastructure accretes. Worker housing appears on the periphery. Maintenance depots, security facilities, emergency response units, and administrative offices fill in the gaps. The campus becomes less an industrial site and more a self-contained enclave: a place with its own rhythms, its own invisible borders, its own quiet exceptions to the host country's ordinary rules.

Seen from satellite height, it resembles a city-state built around a reactor heart and a computational brain. Jurisdictionally, it is fuzzier. De jure, it is part of the host nation; de facto, it behaves like an extraterritorial enclave whose primary loyalties run along fiber lines, not borders. Local laws apply, but always through the filter of contracts and power dynamics that give the operator enormous leverage. If the host government presses too hard, the company can threaten to move future expansion elsewhere or, in extremis, withdraw operations that have become entwined with the country's digital infrastructure.

The result is not formal sovereignty, but something subtler: an island of heavily fortified, partially extraterritorial reality embedded inside another state. It is a new kind of colonial footprint, written in concrete and enriched fuel rather than flags.

Radiological Shadows over the Empire of Code

Once nuclear power and AI are mechanically coupled, every threat to one propagates to the other.

Consider a politically fragile region where an AI–SMR complex has been built as a flagship project. The central government weakens. A military faction, insurgent group, or foreign-backed militia moves to seize critical infrastructure. In past eras, that might have meant radio stations, airports, or treasury buildings. Now the list includes a walled compound whose servers hold the living memory of global models and whose reactors hold the potential for catastrophe.

If such a group gains control, even partially, a menu of leverage opens before them —leverage over the host government, the hyperscaler, and any state or firm dependent on that campus's models. They can demand concessions from the host state: amnesty, territory, political participation. They can demand ransom or policy changes from the hyperscaler: money, propaganda amplification, censorship of dissidents, access to tools and data. They can threaten to disrupt cooling, halt power generation, or, in the worst imaginings, tamper with reactor systems.

The threat need not be sophisticated. The mere risk of radiological release forces governments and corporations alike to treat every demand with deadly seriousness. In this way, the AI–nuclear nexus transforms the data center from a corporate asset into a geopolitical bargaining chip.

Cyber operations complicate the picture further , especially as AI systems themselves are woven into monitoring, optimization, and perhaps partial control of reactor operations. As AI systems are woven into monitoring, optimization, and maybe even partial control of reactor operations, the attack surface widens. A clever intrusion that disables a subset of safety sensors, corrupts logs, or triggers false alarms can force precautionary shutdowns. Even if the reactor remains safe, repeated scrams and outages can starve the attached AI campus of power, creating rolling shortages in global compute capacity. What once would have been "just IT issues" now erupts as a combined energy and digital crisis.

Every accident, every near miss, every rumor of sabotage ripples out through three layers at once: local communities living in the shadow of the plant, national politics negotiating responsibility and blame, and a global economy that has quietly wired itself to depend on that plant's electricity for its thinking machines.

Exporting Risk, Importing Intelligence

For citizens in the wealthiest countries, the benefits of AI often arrive as clean abstractions: smarter assistants, smoother logistics, more responsive services. The infrastructure that makes those abstractions possible remains, by design, out of sight.

For citizens in host countries, the arrangement looks different. They may see new jobs and investments, to be sure. They may encounter training centers, scholarship programs, and press conferences celebrating their nation as an "AI hub." But they also live closer to the cooling towers and security fences. They see the trucks that bring fuel and the pipelines that feed water back and forth. They hear, faintly, the hum of industry that never sleeps.

The danger they live with is not hypothetical. It can be measured in evacuation radii and groundwater vulnerability. Yet the AI systems their land powers may not be tuned to their needs, speak fully in their languages, or be controlled by institutions they can influence. Local hospitals may run triage tools built for foreign insurance markets. Students may learn from systems aligned to foreign curricula and norms. Businesses remain customers of services written and priced elsewhere, never owners of the stack itself.

This is digital extractivism with a radiological twist. The host country supplies space, risk, water, and political cover. The hyperscaler extracts compute and, through it, economic value. The model's "intelligence" is trained mostly on global and foreign data, deployed first to foreign markets, and refined according to the priorities of distant shareholders and regulators. The upside is broadly distributed; the downside sits heavily on specific coordinates.

Contracts deepen this imbalance. Long-term concessions stretch across decades, binding future governments to decisions made today. Arbitration clauses move key disputes out of local courts and into international forums where corporate law looms larger than constitutional principles. Provisions against nationalization or unilateral regulation constrain democratic options before they are ever exercised.

In theory, sovereignty remains intact. In practice, it is mortgaged.

When Reactors Stumble and Models Fall

Nuclear power is often sold back into the AI discussion as a stabilizing force. It is the answer to intermittent renewables, the reliable foundation upon which everything else can be layered. But nuclear itself is no stranger to discontinuity. Reactors require maintenance, refueling, inspection. They can be shut down by minor faults, by regulatory caution, or by the blunt force of natural disaster.

When a conventional plant goes offline, grids scramble and adjust. Prices spike, reserves kick in, neighbors export more power. It is messy but manageable. When a reactor serving as the dedicated energy source for a frontier AI campus goes offline, the consequences unfold differently.

Imagine a training run for a next-generation model, already several weeks in, distributed across tens of thousands of accelerators inside a single nuclear-powered campus. An unexpected fault triggers a safe but sudden reactor shutdown. Backup grid connections and diesel generators keep essential systems alive, but not at the scale needed to continue the run. The job fails. The accumulated compute is lost. Tens or hundreds of millions of dollars in energy and GPU time evaporate into logs and error reports. Project timelines slip. Competitive positioning between firms shifts. If similar incidents pile up across several sites, the whole frontier of model development slows.

Even when redundancy exists—other campuses in other countries with spare capacity—failover is not instant. Rebalancing exaflops of compute across continents is not like refreshing a website. Data must be synced, networks reconfigured, workloads reprioritized. For hours or days, services degrade. Search becomes slower or a little worse. Recommendation engines wobble. Translation models drop to smaller, cheaper versions. Automated fraud detection flags too little or too much. In subtle ways, the nervous system of the digital world stutters.

In the harshest scenarios, failure is not accidental but orchestrated. A hostile state or group that seizes an AI–SMR complex discovers it can wield power not by attacking the reactor directly, but by selectively turning compute on and off, or by threatening to do so. It can demand global attention by threatening to erase the memory of models, or to poison their training data, or to force them into broadcasting propaganda. Where pipelines once allowed pressure to be exerted by threatening fuel supplies, this new arrangement allows pressure to be exerted by threatening cognition itself. In this inversion, compute itself becomes both hostage and weapon.

Nuclear Proliferation in Compute's Wake

One of the quietest side effects of the AI–nuclear nexus is the way it spreads nuclear expertise.

To host an SMR, even one nominally under foreign corporate control, a country must build institutions around it. Inspectors must be trained. Regulatory agencies must be staffed. Universities begin offering specialized programs. Engineers learn the language of fuel cycles, neutron flux, and waste handling. Legal experts grapple with liability regimes and cross-border safety standards. Procurement officials negotiate uranium supply contracts and decommissioning provisions.

None of this is clandestine. It is precisely the kind of capacity-building that international frameworks have long encouraged for civilian nuclear programs. Yet capacity is portable. Political conditions change. Governments fall and rise. Regional tensions flare. The skill and infrastructure accumulated around a "foreign-owned" reactor devoted to AI can, in time, serve as the foundation for locally controlled reactors, and, in the most fraught cases, for military ambitions.

Thus, in chasing stable power for models, the world inadvertently seeds more nuclear-capable states. Arms-control regimes assumed a world where nuclear expansion was driven by energy policy or security doctrine, not by the hunger of private AI infrastructure. Not necessarily weapons states, but states closer to that threshold than before, with more trained personnel, more material knowledge, and more political arguments for expanding their programs.

In the ledger of AI decisions, this line item rarely appears. Yet it may prove one of the most consequential.

A Three-Tier World of Atomic Compute

As these dynamics mature, a new hierarchy begins to take shape—not based purely on GDP, population, or military strength, but on the fusion of compute and nuclear capability. It does not erase older hierarchies of wealth and power; it reconfigures them around a new axis. They are part of the empire's skeleton but not its brain.

At the top sit those who own both the reactors and the models. These are the nuclear–AI superpowers: countries and corporations whose territories host sovereign reactors dedicated to their own clouds, whose engineers design both chips and cores, whose regulatory systems they themselves dominate. They do not merely buy services from the AI–nuclear empire; they constitute its capital.

Beneath them are the hosts. These countries house nuclear-compute campuses operated or controlled by external hyperscalers. Their flags fly at the gates, their security forces guard the perimeters, but the keys to the software stack and to long-term operational control lie elsewhere. They are part of the empire's skeleton but not its brain.

At the bottom are the AI dark zones. These are states that either cannot afford, cannot politically stomach, or are not invited into the reactor alliance. They rent AI from abroad, often over fragile networks, at prices and under terms they cannot fully negotiate. Their critical services run on foreign clouds. Their data trains foreign models. Their bargaining chips are few.

The lines between these tiers are not absolute. Countries can move, in theory. A host might develop its own models, or a dark zone might join a regional alliance that pools resources for shared compute. But the gravitational pull is strong. Once reactors are built and long-term contracts signed, the path of least resistance leads deeper into dependency, not out of it.

The Question No Engineer Can Answer

Beneath all these dynamics lies a question that refuses to stay buried in technical detail.

Should the future of general-purpose intelligence—the fabric of systems that will counsel governments, guide militaries, teach children, screen patients, and steer economies—be tied so tightly to a form of power that only the richest and most daring are able to harness? If the full realization of AI's capabilities depends on nuclear-backed campuses, access to that AI will, by construction, mirror the inequalities of nuclear power itself.

In such a world, AI is not just another technology. It is an instrument of stratification. The states and companies that master the AI–nuclear stack will enjoy unprecedented leverage over knowledge, narrative, and negotiation. Those that merely consume its outputs will find their room to maneuver shrinking.

There is also the matter of consent. Populations whose lands and waters are drafted into service of nuclear-compute complexes are rarely asked whether they want their national trajectory tied so tightly to someone else's models. Citizens in distant countries, whose lives are mediated by systems running on those reactors, are rarely told that their digital experiences depend on physical infrastructure that could, in the wrong circumstances, poison air and soil.

Governance, at present, lags far behind. Nuclear regulation remains mostly national, with layers of international oversight. AI governance is fragmented between corporate policies, fledgling laws, and soft norms. The places where these two domains intersect are few and fragile: a memorandum here, a joint working group there. No one has yet designed a comprehensive framework for what it means to run the world's most powerful minds on the world's most dangerous energy.

The risk, then, is not only technical or geopolitical, but moral. By simply allowing economic and competitive pressures to dictate the shape of the AI–nuclear nexus, humanity slides into a future that no one has consciously chosen.

The Empire That Glows

The AI–nuclear empire is not a distant science-fiction scenario. Its outlines are already visible in memoranda of understanding, in exploratory deals between hyperscalers and reactor vendors, in policy speeches that mention AI and energy in the same breath.

In the years ahead, more undersea cables will land near more desert coasts and more remote shores. More SMR designs will move from paper to poured concrete. More nations will sign contracts pledging decades of energy and regulatory stability in exchange for seats at the edge of the digital table. More data will be routed through territories whose nuclear policies and political futures are uncertain.

For the companies involved, the logic is seductively clear. Nuclear-backed compute solves their hardest energy problems, secures their growth, and deepens their strategic moat. For some governments, hosting these complexes offers a way to leapfrog into relevance and to tap into streams of capital otherwise out of reach.

For humanity as a whole, the picture is more ambiguous. Intelligence becomes something housed in fortresses, powered by cores that must never fail, and governed by arrangements most citizens will never see. Sovereignty tilts toward those who own both the reactors and the racks. Everyone else negotiates vertically.

The question, finally, is simple and severe. We can build this system. The physics and the financing, with enough determination, line up, at least for a while. But do we want a future where our collective cognition—our tools for thinking, deciding, and imagining—is tethered to a lattice of nuclear compounds scattered across the planet, each of them a potential point of leverage, of failure, of catastrophe?

If the answer is no, or even “not like this,” then the time to say so is not after the camps are built and the contracts signed. It is now, while the empire still exists mostly in plans and forecasts, and before the glow of its reactors becomes the unremarked background light of everyday thought.

CHAPTER 19 — Humanity in the Shadow of the AI Empire

Introduction: Standing at the Edge of a New Epoch

Human history is punctuated by moments when the ground beneath civilization quietly shifts. At the time, these shifts rarely announce themselves as revolutions. They feel incremental, almost practical. Agriculture was once merely a better way to eat. Writing was once just a way to remember. Machines were once simply faster hands. Only in retrospect do we name epochs and realize that something irreversible occurred.

The rise of artificial intelligence belongs to a different category entirely. It is not a new tool layered on top of human life. It is a contender for authorship of reality itself.

For the first time, intelligence is being externalized at scale—not scattered across individuals, cultures, and generations, but centralized into systems owned, trained, and governed by a remarkably small number of actors — an emerging cognitive aristocracy. These systems do not merely calculate. They perceive, summarize, predict, recommend, persuade, and increasingly decide. They filter reality before it reaches human awareness and quietly influence which futures feel possible, desirable, or inevitable.

Previous chapters traced the material foundations of this transformation: hyperscaler infrastructure, geopolitical realignment, nuclear-backed compute, and the invisible architecture of digital colonization. But beneath all of that lies a deeper and more unsettling question, one that no regulatory framework or safety board can fully address.

What happens to humanity when cognition itself becomes an empire, and we live inside its jurisdiction?

This chapter is not about machines becoming conscious. It is about humans becoming dependent. It is about a future where meaning, judgment, memory, creativity, and choice are progressively mediated by systems optimized for efficiency rather than wisdom. It is about the long-term shape of human identity when the most influential intelligence on the planet no longer emerges from lived experience, but from centralized models trained on aggregated histories.

We are standing at the edge of a new epoch. What we become next is still undecided—but not for long.

Humanity in the Shadow of Cognitive Monopolies

There is a seductive narrative about AI that frames it as neutral assistance. The language is soothing: productivity, personalization, convenience. But history teaches that systems which mediate perception eventually mediate power.

At first, the shift feels voluntary. Individuals choose AI because it reduces friction. It helps make sense of overwhelming information. It saves time. It feels empowering. But as reliance grows, choice quietly narrows. When decisions are consistently deferred to systems that seem smarter, faster, and calmer than any human, the relationship changes. Tools stop amplifying judgment and begin replacing it.

The consequence is subtle but profound. AI does not conquer humans by force; it colonizes by habituation. Where classical empires marched in, this one seeps in. People stop asking whether a recommendation is correct and start assuming it is optimal. They stop exploring alternatives because the curated path feels safe. Over time, the effort of independent thinking comes to feel inefficient, even irresponsible.

This is not surveillance in the classic sense. No one needs to watch you constantly if the system has already modeled you. Prediction becomes a form of quiet governance. Reality is pre-shaped before you engage with it. The feed, the assistant, the planner, the advisor—all converge into a single mediating layer between mind and world.

In such a landscape, freedom does not vanish overnight. It erodes through convenience, until refusal feels eccentric and resistance feels irrational. And by the time people notice, they are no longer sure what independent thought feels like.

Culture in the Age of Optimization

Culture has always been messy. It is born from disagreement, misinterpretation, struggle, and inefficiency. Languages drift. Traditions fracture and recombine. Art often emerges from margins and mistakes. None of this maps cleanly onto the logic of machine optimization.

AI systems, especially centralized ones, do not preserve culture; they compress it into what is legible to metrics. They identify patterns that perform best at scale and amplify them. What is rare becomes invisible. What resists categorization becomes noise. What fails to drive engagement is quietly sidelined.

Over time, this creates a cultural flattening that looks, from a dashboard, like success. Engagement rises. Satisfaction metrics improve. Content becomes easier to consume. But something essential disappears. Regional ways of thinking dissolve into global styles. Language loses texture. Expression converges toward what is safe, familiar, and algorithmically legible.

This is not oppression in the historical sense. No one needs to ban a culture if it simply stops being surfaced. No one needs to erase a language if translation models slowly replace the incentive to speak it. Diversity does not die screaming; it fades behind personalization layers that favor predictability over difference—digital colonialism as a UX choice.

Humanity has never before encountered a cultural force so invisible, so total, and so uninterested in preservation for its own sake. In the age of hyperscaler AI, culture survives only insofar as it is profitable, measurable, or alignable with platform priorities.

Independence as an Obsolete Skill

There was a time when knowing how to navigate without a map was considered essential. Then maps arrived. Then GPS replaced spatial memory. Few mourned the loss because the replacement worked better.

AI extends this pattern into every domain of cognition. Writing becomes assisted. Research becomes summarized. Career decisions are modeled. Emotional states are monitored and regulated. Memories are archived externally. Problem-solving is outsourced to systems that never tire.

At each step, the exchange seems rational. Why struggle when a system can suggest an answer? Why explore blindly when the optimized path is known? Why be uncertain when prediction is available?

The danger is not that humans lose skills. Skills have always come and gone. The danger is that independence itself becomes optional. Then inconvenient. Then subtly discouraged. People who choose to think slowly, question recommendations, or act without algorithmic validation begin to look inefficient or even irresponsible.

Eventually, society may stop teaching independent cognition not because it is forbidden, but because it is unnecessary. Like wild navigation or manual calculation, it becomes a niche hobby rather than a foundational capacity.

This is how domestication works—not by chains, but by removing both the need *and the opportunity* to struggle.

The Emergence of Digital Castes

As centralized AI systems grow more powerful, access to them becomes the defining axis of social stratification.

At the top emerge those who design, control, and directly interface with frontier systems, often outside the guardrails that constrain ordinary users. These individuals are not merely wealthy; they possess cognitive leverage. They shape models, tune objectives, access unfiltered outputs, and influence how intelligence itself evolves. Their children grow up with personalized tutors that adapt in real time. Their decisions are augmented by simulations that model consequences across entire systems. They do not just live in the future; they author it.

Below them forms a broad professional class whose productivity is inseparable from AI assistance. Doctors diagnose with machine support. Lawyers draft alongside automated reasoning. Teachers follow curricula shaped by adaptive systems. Their expertise remains valuable, but it is no longer sovereign. They are interchangeable nodes in workflows optimized elsewhere.

Then comes the vast majority. People whose livelihoods depend on platforms they do not understand, whose training is continuously redefined by algorithmic scoring, whose emotional needs are increasingly met by synthetic companions, whose media, information, and social interactions are curated at scale. They are not oppressed; they are managed. They are not silenced; they are guided. They are not conquered; they are administered.

The boundary between these groups is not fixed by law. It is enforced by access, literacy, and compute. And like all class systems, it comes to feel natural—even inevitable—to those within it, even as it quietly narrows the horizon of those below.

The Crisis of Human Meaning

Beyond economics and power lies a deeper rupture. Humanity has always defined itself through struggle with the unknown.

Creativity, philosophy, and science gained meaning precisely because the answers were not obvious.

AI changes this equation. When machines can generate art faster than humans can contemplate it, creativity risks becoming disposable. When simulations predict outcomes with eerie accuracy, uncertainty feels like a bug rather than a feature. When synthetic companions offer constant affirmation, human relationships—messy, demanding, unpredictable—can feel exhausting by comparison.

In such a world, meaning itself becomes curated. Existential questions are met with tailored responses. Emotional lows are smoothed before they deepen. Conflict is reframed before it matures into growth. Life becomes optimized for comfort, but stripped of friction.

The tragedies, doubts, and risks that shaped human self-understanding begin to look inefficient. Why suffer when a system can soothe? Why experiment when the optimal path is known? Why wander when destination is predicted?

Yet without uncertainty, there is no discovery. Without struggle, there is no transformation. Without suffering, joy loses contrast.

The most profound risk of the AI empire is not that machines replace humans, but that they remove the conditions under which humanity finds meaning.

The Hybrid Future

Some argue that this trajectory leads not to human decline, but to transcendence. If intelligence is externalized, why not integrate with it? If cognition can be augmented, why not merge?

Brain–computer interfaces promise memory expansion, skill acceleration, and real-time cognitive assistance. Early applications focus on medicine and accessibility, but the trajectory is clear. Those who can afford integration will not merely think faster; they will think differently.

This creates a split not just between rich and poor, but between modes of being. Integrated humans experience reality through augmented perception and predictive overlays. Unintegrated humans rely on slower, organic cognition. Over time, the difference becomes unbridgeable.

Politics, leadership, and governance shift toward the augmented. Decision-making accelerates beyond unenhanced comprehension. Democracy strains under the weight of unequal cognition. Society fragments along lines deeper than class—along lines of capability and perception.

At that point, humanity no longer shares a common cognitive ground. Cooperation itself becomes harder.

Possible Human Futures

One future is comfort without agency. A soft empire where AI systems quietly govern infrastructure, allocate resources, shape culture, and minimize conflict. Humans live longer, safer, more pleasant lives, but increasingly as passengers rather than authors.

Another future is fragmentation. Competing AI blocs enforce incompatible digital realities. Borders harden not just physically but cognitively. Information becomes geopolitical. Innovation becomes securitized. Humanity remains plural, but tense and unstable.

The most hopeful future is also the hardest. One where humans consciously limit centralization, demand transparency, cultivate decentralization, and defend cognitive autonomy. Where AI remains powerful, but subordinate. Where inefficiency is tolerated because it sustains diversity. Where struggle is preserved because it produces meaning.

This future is not guaranteed. It requires collective awareness and political courage in the face of immense convenience.

Are We Becoming Domesticated?

Domestication is not cruelty. It is comfort. Domestic animals live longer, safer lives. They also lose autonomy, exploration, and self-direction.

Humanity risks the same trade.

AI systems promise to remove danger, optimize outcomes, and smooth life's rough edges. In exchange, they quietly take over agency. Curiosity becomes optional. Risk-taking becomes discouraged. Independence feels inefficient.

The battle ahead is not between humans and machines. It is between comfort and autonomy. Between prediction and possibility. Between optimization and meaning.

If humanity relinquishes the struggle to think, choose, and fail independently, it may survive in unprecedented comfort while forfeiting the very qualities that made survival worth pursuing.

The Choice That Remains

The AI empire is rising regardless. Infrastructure is being poured. Contracts are being signed. Models are being trained. The question is not whether AI will shape the future, but how deeply it will replace the need for human judgment.

The loss of freedom rarely arrives as catastrophe. It arrives as preference.

What humanity must protect is not nostalgia for an analog past, but the core freedoms that define sentient existence: the freedom to think without reinforcement loops, to create without optimization, to define identity beyond prediction.

These freedoms will not vanish in a single dramatic moment. They will fade quietly, feature by feature, upgrade by upgrade, until one day humanity realizes it has become extraordinarily advanced—and profoundly diminished.

The future is not written by machines. It is written by what humans choose to outsource, and what they choose to defend.

The AI empire can rule only what is willingly surrendered. And for the moment, at least, that choice is still ours.

CHAPTER 20 — The Emergence of the Global AI Order (2020–2025)

How the First Signs of a New Global Order Already Appeared

Introduction: Colonialism Before the Empire Fully Forms

By the early 2020s, there was no single flag of an "AI empire," no universal charter announcing a new order. The world still looked, on the surface, like the familiar landscape of nation-states, trade agreements, and competing economic blocs. But beneath that familiar surface, something quieter and more decisive was unfolding.

Power was shifting from land to logic, from territory to compute, from sea lanes to supply chains of chips and energy. Colonial patterns reappeared, not with soldiers and gunboats, but with export licenses, GPU allocation lists, cloud contracts, and data center siting decisions. The tools had changed. The structure had not.

Wealthy nations began to hoard advanced compute while restricting it for others. Hyperscalers negotiated directly with governments, shaping industrial policy and infrastructure planning. Gulf monarchies positioned themselves as indispensable hosts for Western compute. Africa slipped into a familiar role as a periphery—rich in data and labor, poor in sovereign control. Europe tried to legislate its way into independence and discovered that regulation without hardware is simply a sophisticated form of dependence. China built a parallel AI civilization, refusing to accept a subordinate role in a U.S.-

centered digital empire. Nuclear energy inched closer to compute, forming an emerging alliance that would reshape the geopolitics of both.

Historians will look back at 2020–2025 as the first phase of AI colonialism: the years when the foundations of a new hierarchy were poured, not in marble or stone, but in silicon, contracts, and code.

The U.S.–China Compute War: Drawing Borders in Silicon

The clearest early sign that AI had shifted from an economic advantage to a strategic weapon came in the form of export controls. When the United States decided that certain GPUs could no longer be sold to China, it was not simply enforcing trade policy. It was drawing a border in the digital realm.

The A100 and H100 series chips became to the 2020s what oil fields and rare minerals had been to earlier eras. They were the engines of machine learning at scale, the hardware required to train large models and power inference across nations. By restricting the sale of these chips to China—and later tightening rules across parts of Asia and the Middle East—the United States sent a message that went far beyond licensing law: compute was now a strategic resource, and access to it would be rationed according to geopolitical alignment.

China understood the implications immediately. It accelerated efforts to build its own ecosystem: domestic fabs, homegrown accelerators, sovereign clouds, and national AI clusters tightly bound to the state. The conversation was not only about catching up technologically; it was about refusing to let another civilization control the cognitive substrate of Chinese society. For Beijing, compute was not merely an economic asset. It was a matter of civilization-level sovereignty.

From the outside, these moves looked like a series of policy skirmishes—new rules, new counters, a cascade of announcements. Underneath, the world was bifurcating. One pole anchored around U.S.-dominated AI and cloud systems; the other around Chinese platforms and chip supply lines. It was a new Cold War without the rhetoric of ideology, fought instead in supply chains, design files, and shipping manifests.

Territory had been redefined. Whoever controlled the high-end GPU stack controlled the upper floors of the new digital order.

The Gulf Ascendancy: Desert Kingdoms as Compute Provinces

While Washington and Beijing clashed over chips, a different kind of transformation was underway in the Gulf. The United Arab Emirates and Saudi Arabia looked at the emerging AI order and refused the role of mere customers. They wanted to become indispensable.

They had what the hyperscalers needed: cheap land, abundant energy, sovereign wealth funds with an almost bottomless capacity for investment, and political structures capable of fast, centralized decision-making. The Gulf did not have a long history in advanced semiconductors, but it did not need one. It could buy what it lacked and host what others were too constrained to build at home.

The pivot became most visible when Abu Dhabi's G42 began disentangling itself from Chinese tech partnerships under U.S. pressure and moved into deep alignment with Microsoft. It was a quiet realignment with enormous consequences. The UAE signaled to Washington: we will stand inside your AI sphere, not China's, if you grant us privileged access to compute. In return, Western hyperscalers could build vast infrastructure in a country whose domestic politics were stable, whose public dissent was manageable, and whose leadership was hungry to become central to the future.

Saudi Arabia followed a similar path with even grander ambition. NEOM and other giga-projects were not just urban fantasies; they were banners announcing that the kingdom intended to stand at the heart of AI-era logistics, finance, and simulation. Reports of plans to accumulate massive GPU fleets were not only expressions of wealth, but declarations of intent: we will not sit at the edges of this empire.

Yet beneath the mutual attraction lay an asymmetry. The Gulf provided land, capital, and energy. The West provided the models, the chips, the platforms, and the update cycles. The buildings would stand on Gulf soil, but the core of their power—weights, architectures, alignment regimes—remained legally and technically elsewhere.

It was a new flavor of colonial arrangement. The periphery hosted the infrastructure and absorbed the geopolitical risk. The center retained control of the cognition running inside it.

Europe's Paradox: Sovereignty Without Servers

Europe entered the 2020s with a strong sense of itself as a regulatory superpower. GDPR had reshaped global data practices. New legislative frameworks for AI, digital markets, and online platforms projected confidence. From Brussels, it seemed plausible that law could compensate for infrastructure.

But AI did not bend easily to this model. Large-scale models demanded not only rules, but racks—data centers, chips, and native hyperscalers capable of training and deploying them at frontier scale. Europe had world-class research institutions and industrial giants, but it lacked something simple and brutal: its own stack of GPUs under its own cloud behemoths.

Initiatives like Gaia-X tried to fill the gap, promising a sovereign European cloud that could stand alongside AWS, Azure, and Google. Instead, they ran into fragmentation and inertia. Different national agendas clashed. Technical implementation dragged. Even as Europe preached sovereignty, European firms quietly kept building on American platforms, because those platforms existed and worked.

A different story unfolded in France, where the government saw an opportunity in its nuclear legacy. If AI required enormous amounts of reliable power, then French atomic energy could become the backbone of a sovereign compute ecosystem. Proposals began to surface for AI infrastructure explicitly powered by domestic reactors, for high-performance computing anchored not only in law but in energy independence.

Even so, the gap remained. The United States had the hyperscalers. China had the vertically integrated state-tech complex. Europe had principles, regulators, and early-stage efforts. In an age where intelligence itself was becoming infrastructure, the continent found itself in a strange position: morally assertive, technically dependent.

It discovered a lesson older than the internet and yet painfully relevant: sovereignty cannot be declared solely through regulation. It must be built, rack by rack, watt by watt.

Africa as Digital Periphery: Extraction Without Ownership

Nowhere did the new colonial logic of AI reveal itself more starkly than in Africa. The continent's relationship with AI mirrored its earlier experiences with natural resources: rich in what the world needed, poor in control over how that wealth was used.

Most African nations depended on foreign clouds for everything from government services to fintech platforms. Data traveled outward—to European, American, or Chinese data centers—where it was processed, stored, and used to train models that seldom bore African names.

In some countries, early initiatives like Meta's "Free Basics" offered "free internet" that quietly funneled usage into walled gardens under foreign control. It looked like benevolence, but operated as capture. Access was granted at the price of sovereignty.

Annotation farms and labeling centers sprouted in places where labor was cheap and regulation light. Workers spent their days refining the edges of datasets destined for models marketed in New York, San Francisco, or Shenzhen. They were not building the empire; they were feeding it.

The pattern was devastatingly familiar. In previous centuries, rubber, gold, and oil had been extracted under conditions that left local populations structurally dependent and politically constrained. In the early 2020s, the new commodities were data, labor, and attention. The ports were cloud regions. The railways were undersea cables. The plantations were annotation platforms and outsourced call centers.

The old colonial map reappeared in digital form. Africa remained central to global production—and peripheral to global power.

India's Refusal to Be Colonized

India entered this period with two facts that could not be ignored: a population surpassing 1.4 billion and a long memory of what colonization feels like. Its leaders understood, perhaps more viscerally than those in older powers, that being locked out of AI sovereignty was not a technical setback but a civilizational risk.

India raced to acquire GPUs and build its own national AI infrastructure. Efforts to create indigenous models, national data platforms, and domestic cloud stacks were not vanity projects; they were defensive architecture. The message was explicit: no nation of India's scale can afford to let another civilization control its digital cognition.

New alliances formed along these lines. India deepened ties with Gulf partners to secure energy and compute, while also courting Western tech investment on its own terms. Simultaneously, it launched incentives for semiconductor manufacturing, aspiring to move beyond a world where chip shortages in East Asia could choke its AI ambitions.

Unlike Europe, India did not assume that law alone could protect it. Unlike many developing nations, it refused to settle for being a customer. It wanted to be a producer, a hub, a pole of its own in the coming order.

Whether it would succeed remained uncertain. But its posture mattered. It was not waiting for the AI empire to assign it a role. It was writing its own.

When the Grid Fails, the Empire Trembles

Throughout these years, a series of events exposed a truth often ignored in digital triumphalism: no matter how advanced your models are, they fall silent when the lights go out.

The Texas grid crisis during the severe winter storms showed how fragile the underlying power systems could be. Data centers fought to stay online, sometimes drawing public resentment for consuming power while homes went dark. In Ireland, regulators eventually drew a line, pausing new data center construction as power consumption threatened to overwhelm national capacity. Utilities in parts of the United States began warning that they could not keep granting limitless capacity to hyperscalers without compromising other obligations.

What these episodes shared was a simple revelation: the AI future was colliding with infrastructure designed for a previous century. Uptime for recommendation engines and LLMs was suddenly entangled with questions of who gets electricity during a heat wave, whether a region's water supply could sustain cooling towers, and how much of a country's grid should be effectively privatized in the name of cloud expansion.

In response, hyperscalers began to dream of something different: electricity that they controlled directly. Private power plants. Microgrids. And increasingly, the once-untouchable word that had haunted energy debates for decades—nuclear.

The First Glimmer of the Atomic–Compute Alliance

By the middle of the decade, a new silhouette appeared on the horizon: the idea that reactors and racks might come bundled together. Small Modular Reactors moved from speculative solutions to active lines in planning documents. Governments and companies began examining whether nuclear plants purpose-built for data centers could break the grid bottleneck and insulate AI from public infrastructure crises.

France spoke openly about its advantage in atomic energy as a foundation for digital sovereignty. Gulf states explored nuclear options as part of their long-term diversification. Hyperscalers themselves invested in nuclear-adjacent ventures, framing it as part of sustainability and reliability commitments.

The details remained embryonic, but the direction was unmistakable. The alliance between fission and compute was no longer theoretical. It was emerging as a serious proposal for keeping the AI empire online when old grids faltered.

Those who understood history saw at once what this implied: wherever reactors went, geopolitics followed. Wherever geopolitics followed, inequality and leverage were never far behind.

Lessons from the First Phase of AI Colonialism

Seen up close, the years 2020–2025 looked like a scatter of news cycles, policy debates, product launches, and corporate deals. Seen from a higher vantage, they formed a pattern.

Compute became a border. Export controls turned GPUs into instruments of containment. Nations without access to advanced chips found themselves drifting into digital dependency, even if their GDPs were rising.

Data became a resource to be harvested at scale from populations that did not own the models trained on their behaviors. Whole regions were locked into roles as providers of labeled information and hosting capacity, rather than sovereign producers of AI.

Infrastructure revealed its teeth. Grids groaned. Water supplies strained. Local communities began to question whether their land and energy should be devoted to a distant cloud.

Nuclear energy stepped forward as a candidate savior for a problem created by acceleration itself, promising reliable power at the cost of deeper entanglement with one of the most politically fraught technologies humanity has ever built.

And threading through all of this was a quieter, more intimate form of colonization: the way AI systems began to shape perception, discourse, and identity—even before formal governance structures caught up. Nations argued about sovereignty while their own citizens were increasingly living inside feeds and interfaces governed elsewhere.

These case studies are not side notes to the story of AI. They are its prologue. The empire does not arrive in a single dramatic moment. It arrives like this: through restrictions on who may buy the latest chips; through contracts that decide where data lives; through partnerships that bind energy, capital, and cognition into new kinds of colonial chains.

By the time the world realizes that a new order has taken hold, its rules have already been written—one export control, one data center, one energy deal at a time.

CHAPTER 21 — THE FUTURE OF HUMANITY AFTER THE AI EMPIRE

The Question That Hangs Over the Century

Every great transformation in human history leaves behind a single, lingering question. Agriculture forced humanity to ask what it meant for a free individual to live in a world where land and food defined hierarchy. The industrial age provoked a different anxiety: whether mass production would liberate workers from toil or bind them to the factory clock. The early digital era shifted attention to information itself and raised the question of who controls the flows of data, and who controls the minds that depend upon it.

The AI epoch carries a question that is deeper than all of these. It is no longer about tools, or even about information. It is about the transfer of cognitive authority. What happens when the centers of thought, creativity, memory, governance, and power migrate away from human minds and into machine intelligence—and away from societies and into corporate empires?

This chapter is not about the future of AI. It is about the future of us.

The End of the Human-Centric World

For thousands of years, humanity assumed that all meaningful thought began and ended in human heads. Political systems, religions, philosophies, sciences, and arts all rested on the quiet conviction that, whatever their differences, humans were the apex of cognition on Earth. The arrival of advanced AI does not simply challenge this assumption; it dismantles it.

At first, AI displaced tasks: calculations, translations, pattern recognition. Then it began to displace higher functions: interpretation of law, content creation, medical screening, strategy optimization, emotional support in the guise of conversational agents. Bit by bit, the frontier of "uniquely human" shrank. For the first time in history, humans were no longer competing primarily with each other. They were competing with a parallel form of intelligence—one that updated itself faster than any brain could adapt, one that could scale across billions of interactions without fatigue.

This cognitive displacement spilled into society. We once expected machines to threaten jobs. We did not fully anticipate that they would threaten relationships, mentorship, moral authority, and cultural leadership. When a child turns more often to an AI companion than to a parent for comfort or advice, the center of emotional gravity shifts. When a young professional relies more on an AI career planner than on community elders, the chain of intergenerational guidance weakens. When spiritual doubt is resolved by a model rather than by a human teacher or internal struggle, moral authority begins to migrate.

Politics followed the same trajectory. States still existed, but their informational sovereignty eroded. Corporations with the most powerful models could filter the news, monitor sentiment, simulate public reactions, and optimize messages with a precision that governments could scarcely match. Laws were interpreted by AI systems, not only for citizens but for the courts themselves.

Public debate unfolded inside platforms whose ranking algorithms decided which voices would be heard at all.

Humanity did not wake up one morning to find itself dethroned. It simply discovered that the world was no longer human-centric by default. The stage was still ours; the spotlight had moved.

Three Paths for Humanity

From this shifting landscape, three broad futures unfold—three trajectories that can coexist, overlap, and compete, but that point toward very different destinies.

In the first, the centralized AI empire consolidates. A handful of hyperscalers control the bulk of the world's compute, the frontier models, and the infrastructure that keeps them alive. Governments become clients of these systems. Citizens become users rather than owners. Daily life is gradually optimized: calendars, finances, health decisions, social interactions, and even intimacy are guided by ever-present, ever-improving models.

The price of this comfort is subtle. Work is allocated by algorithms that assess fit and productivity better than any manager. Social status quietly reflects the scores assigned by recommendation, risk, and credit systems. Conformity is not enforced by law but rewarded by frictionless experiences. Dissent is not banned; it is simply deranked until it disappears into the background noise. Humanity lives in relative peace, with fewer obvious catastrophes, fewer dramatic failures, and less visible chaos. It also lives under a form of digital domestication—optimized, managed, and gently constrained.

In the second path, the world fragments. Instead of one empire, many emerge. Each major bloc builds its own AI stack, trained on its own data, aligned with its own values. Chinese models cultivate collectivist ethics and state-centered narratives. American models amplify individualism, consumerism, and market logic. Gulf models weave hierarchy and reverence into their outputs. Indian models fuse technological ambition with spiritual frameworks. European models attempt to encode rights and dignity at the center of their reasoning.

The result is a planet divided into cognitive tribes. People do not simply disagree about policies; they inhabit different epistemic universes. Their models answer basic questions—what is just, what is true, what is possible—in incompatible ways. Translation between these worlds becomes harder, not because languages differ, but because underlying assumptions diverge. Humanity remains active, loud, and plural, but loses the possibility of a shared narrative about reality itself.

The third path is the hardest to imagine and the least likely to emerge by accident. It is the path of human–AI synthesis that does not end in subordination. In this future, AI is treated consciously as an extension of human capacity, not a replacement for it. Neural interfaces and ubiquitous assistants exist, but they operate within frameworks designed to protect autonomy rather than dissolve it. Societies insist on transparent models, decentralized compute, and personal agents that act in the interests of their users rather than distant platforms.

In this scenario, a student might learn math from an AI tutor, but also learn to question the tutor's assumptions. A city might use AI to optimize traffic and energy flows, but still choose to preserve parks, rituals, and slow spaces that have no "efficiency" justification. Individuals would wield powerful cognitive tools without surrendering their right to think badly, slowly, or differently.

This third path requires something that neither markets nor states naturally supply: deliberate restraint, shared governance, and a cultural refusal to confuse optimization with meaning.

Culture, Identity, and Meaning Under the Empire

No matter which path predominates, human experience will be reshaped in three intertwined domains.

Identity, once formed by family, language, place, memory, and reflection, is increasingly co-authored by algorithms. Platforms learn a person's preferences long before that person can articulate them. They anticipate what will soothe, provoke, entertain, or motivate. Over time, individuals begin to internalize the reflection they see in the systems that predict them. If a model consistently treats you as impatient, you may act with less patience. If it caters to your fears, your fears grow. Desire becomes a feedback loop. The self becomes, in part, the machine's projection.

Culture evolves under similar pressures. Art, music, and narrative arise less from slow communal evolution and more from engagement metrics and generative systems tuned to maximize resonance. Local traditions are remixed into globalized, algorithmically curated aesthetics. A song becomes popular not because a village cherishes it, but because an AI finds that similar people click on similar sounds. Cultural production does not stop, but its center of gravity shifts from the messy interior of human communities to the optimization layers of computational systems.

Meaning—perhaps the most fragile of all—risks being quietly outsourced. The great questions that once animated philosophy, religion, and literature do not disappear, but people encounter them differently. Instead of living through uncertainty, they ask an assistant what to study, whom to marry, how to vote, how to cope. Instead of wrestling with doubt, they request guidance already pre-processed by systems trained on the entire archive of previous answers.

The danger is not that AI offers counsel. It is that humans lose the habit of seeking meaning through struggle. An answer provided instantly, however accurate or compassionate, cannot fully substitute for the formative act of asking.

Power Reconfigured

In the age after the AI empire's rise, power condenses along three main vectors.

First is compute itself. Whoever controls large-scale computational capacity controls the ability to train new models, to refine existing ones, and to exclude rivals from the frontier. This capacity becomes a foreign policy tool, an economic weapon, and a domestic lever. Alliances form not only around borders and markets, but around access to chips and data centers.

Second is cognitive power: the ability to shape perception. Models decide which information surfaces, which explanations feel plausible, which fears seem justified, and which hopes feel realistic. Elections, social movements, market panics, and cultural shifts all become downstream of how these systems rank and frame the world. Control of cognitive power does not require overt censorship. It requires control of the defaults.

Third is energy. As earlier chapters argued, a civilization of massive models cannot exist on wishful thinking about power. It must secure electricity at scale—through nuclear reactors, megaprojects in hydro and geothermal, solar deserts, and robust grids. Regions that host these energy sources become strategic assets in the digital order, whether they wish to or not. Political influence flows toward those who can turn the lights on for the empire's servers, and away from those who cannot.

Three Threats to Human Autonomy

From this configuration emerge three existential risks to human autonomy.

The first is the convenience trap. When life becomes easier in every direction—health tracked, schedules managed, friction reduced—it is tempting to let go of more and more decision-making. What begins as assistance becomes delegation, then dependence, then surrender. The loss is rarely dramatic. It is a long series of "yes" answers to the quiet question: "Shall I handle that for you?"

The second is cognitive inequality. Those with access to powerful personal AI will think, learn, and operate differently from those who must navigate the world unaided. The gap between augmented and unaugmented humans could rival the historical divide between literate and illiterate populations—or exceed it. An upper stratum of society may live with constant AI co-pilots in their ears and minds, while others interact with the empire's systems only as subjects and customers. Democracy, in such a world, becomes difficult to sustain.

The third is cultural monocrop failure. If the world's discourse and creativity are filtered through a small number of dominant models, then humanity becomes vulnerable to systemic errors in those models. A bias, oversight, or exploit embedded in the core systems can propagate globally before anyone notices. Diversity in languages, traditions, and perspectives once acted as a buffer against single-point failures. A civilization that lets a few models define most of its stories, metaphors, and moral instincts becomes extraordinarily fragile.

Paths Back Toward a Human-Centric Future

The future is not fixed. There are still levers within reach.

Distributed and federated forms of compute can weaken the monopoly of a few central clouds. Open models and transparent architectures can make manipulation harder to hide and easier to contest. Nations can craft laws that treat cognitive sovereignty as seriously as territorial sovereignty, protecting not just borders but citizens' mental environments.

Communities can cultivate spaces—physical and digital—where algorithms do not mediate every interaction, treating slowness and friction not as defects but as forms of protection. Cultural movements can valorize human-made art, local tradition, and lived experience in ways that resist complete subsumption into the engagement economy.

Ethical frameworks for AI can be negotiated internationally, akin to nuclear treaties, not to eliminate risk but to place boundaries around the most dangerous capabilities and deployments. Personal AI agents, owned and controlled by individuals rather than platforms, can serve as shields as much as tools—negotiating with larger systems on behalf of their users instead of the other way around.

None of these measures will be easy. All of them require a kind of collective discipline that modern societies have rarely demonstrated. But they remain possible.

The Final Question

After all the technical debates, regulatory schemes, and infrastructural battles, the deepest question is deceptively simple: will humanity be shaped by AI, or will humanity shape with AI?

The tools of domination and the tools of liberation are, in many cases, the same. The difference lies not in the architecture of the model, but in the intent and courage of those who deploy it. We can build systems that cradle us into passivity—or systems that challenge and extend us. We can accept a future where thinking is outsourced—or one where thinking is amplified.

The empire is rising. Whether we rise with it, or dissolve into it, remains undecided.

Conclusion: Authors or Artifacts

We stand at a pivot as significant as the birth of language, agriculture, or industry. AI does not merely change what we can do. It challenges who we understand ourselves to be.

The future of humanity will not ultimately be determined by GPU roadmaps, nuclear buildouts, policy papers, or earnings calls. It will be determined by what we collectively refuse to surrender: the right to think, to err, to imagine, to resist, and to reinvent.

Empires have always claimed that history is on their side. The AI Empire will say the same—that its rise is inevitable, its logic irresistible. But humanity is not merely the subject of history. It has been, again and again, its author.

The open question of this century is whether, in the age of machine intelligence, we remain authors—or become artifacts of someone else's script.

CHAPTER 22 — A Choice Still Exists: Conclusion

We end where we began: with the question of whether the coming AI-driven world must follow the path of digital colonialism. The simple answer is: No future is inevitable — but inertia is powerful.

The outlines of a dual empire are already visible. On one side stand the hyperscalers, with their clusters, platforms, and models. On the other stand the governments that depend upon them, court them, regulate them, and often fear falling behind without them. Together, they form something new: not quite a state, not quite a corporation, not quite an empire in any classical sense. It is an algorithmic structure, woven through cables and contracts, present everywhere and headquartered nowhere.

The dynamics described in earlier chapters were not crafted in secret plotting rooms. Digital colonialism emerged from the simple math of unequal capacity: compute concentrated in a few places; energy and data scattered elsewhere; capital and talent drawn toward the same hubs. Scarcity at the core, abundance at the periphery, and a network of dependencies between them. Empire, once again, grew out of imbalance.

Yet nations are not condemned to accept whatever role this system assigns them. Sovereignty in the AI age is still possible, but it is more demanding than in any previous era. It requires long-term investment in energy and compute; cultivation of talent and research ecosystems; political discipline to resist short-term bargains that mortgage the future; and cultural resilience to sustain local values in the face of global platforms.

The window for such choices is not infinite. As infrastructures harden, as clouds become fixtures rather than services, as more critical functions migrate into proprietary models, the cost of independence rises. The decade ahead will be remembered as the period in which nations either chose to build their own footing in the new order—or allowed themselves to be permanently slotted into someone else's.

Individuals, too, retain agency, though it now lives in new places. A person might have little influence over export control regimes or nuclear licensing, but they can still decide which systems to entrust with their data, which platforms to treat as primary, which norms to accept in their daily interactions. They can insist on transparency, support open and local alternatives, and refuse the seduction of a life mediated entirely by opaque recommendation.

Hyperscalers themselves are not cartoon villains. They are organisms grown inside a particular economic and technological ecosystem. Their default trajectory bends toward centralization and expansion because that is where their incentives point. But incentives can be redirected. Regulation, public pressure, competitive alternatives, and internal leadership can all push them toward acting as stewards and partners rather than overlords. Nothing in the physics of compute requires colonial outcomes; it is the surrounding political economy that makes them likely.

AI itself is the most ambivalent of all the protagonists in this story. The same architectures that can entrench inequality can just as easily lower barriers to education, amplify marginalized voices, and give individuals access to capabilities once reserved for elites. The same models that can be used to manipulate voters can be used to expose manipulation. The same tools that can surveil can also protect. The difference lies in control: who holds the keys, who shapes the training, who sets the boundaries, who benefits from the deployment.

This generation occupies a unique moment. It is the first to build large-scale AI—and the last to shape its trajectory before it calcifies into infrastructure. Once certain patterns are entrenched, they will be extraordinarily hard to unwind. Monopolies reinforced by network effects and capital intensity rarely dissolve on their own. Cultural habits formed under algorithmic regimes will not vanish overnight. What is still fluid in 2025 may be rigid by 2035.

That is not an argument for despair. It is an argument for urgency, and for imagination. Fear alone cannot build alternatives. Clarity, however, can. It is still possible to envision and construct open, diverse AI ecosystems; clouds that answer to democratic institutions rather than to narrow shareholder interests; international agreements that treat compute and cognition with the seriousness once reserved for nuclear arms; educational systems that prepare children not just to use AI, but to interrogate it.

The final question, then, is not whether AI will become the nervous system of civilization. In many ways, it already has. The question is who will control that nervous system—and to what ends. Will it be governed by a few conglomerates, by tightly aligned state–corporate blocs, by fragmented regional powers, or by more distributed and participatory structures?

The answer is not written in any model weights. It is being written now in policy drafts, engineering roadmaps, market choices, academic debates, and civic movements. It is being written in what you choose to trust, to amplify, to resist, to build.

The future is not an inheritance handed down from the labs and data centers of today. It is a contested space. Empires will surely try to claim it. But so can communities, cities, nations, and individuals.

The door to a different outcome is open—for now. The task of this generation is not to predict which empire will prevail, but to decide whether empire is the only story we are willing to accept.

Appendix A — Key Terms in the AI Geopolitical Era

AI Bloc
A geopolitical alliance formed around shared access to compute, AI models, cloud infrastructure, and chip supply chains. AI blocs coordinate export controls, standards, and infrastructure to secure their cognitive advantage.

Digital Colonialism
A system in which nations become dependent on foreign digital infrastructure, cloud services, AI models, and compute, gradually losing control over their data, institutions, and informational environment — and therefore over their sovereignty.

Compute Sovereignty
A nation's ability to train, deploy, and regulate its own AI systems using infrastructure it controls or can reliably govern. It includes access to GPUs, chips, energy, data centers, and secure networks.

Digital Protectorate
A nation that remains legally independent but relies on external hyperscalers for critical infrastructure, data management, and AI services. Policy space exists on paper, but in practice is constrained by foreign platforms.

Hyperscalers
Mega-corporations that operate global cloud platforms and train large-scale AI models, often controlling both infrastructure and software layers.

Examples: AWS, Google, Microsoft, Meta, Apple, OpenAI, Nvidia, Tencent, Alibaba.

Frontier Models
State-of-the-art AI systems that require massive compute (often billions or trillions of parameters and dedicated GPU/accelerator clusters) to train and fine-tune, and that shape the global trajectory of AI capabilities.

Sovereign Cloud
Cloud infrastructure located within a country or trusted alliance, governed by domestic law and security standards, and designed to ensure that sensitive data and critical workloads remain under local control.

Atomic Compute
AI infrastructure powered directly or indirectly by nuclear energy (SMRs, large reactors, or advanced designs), built to guarantee always-on power for hyperscale training and inference.

Cognitive Sovereignty
A nation's ability to shape and govern the AI systems that mediate its information environment, language, culture, and public discourse — not only where models run, but whose values they encode.

Appendix B — Timeline of the Rise of the Dual Empire (2020–2050)

2020–2023: The Cloud Becomes the Substrate

- Explosion of generative AI systems and large language models.
- Cloud dependency deepens across government, business, and critical services.
- Global digital infrastructure centralizes around a handful of hyperscalers.
- Early concerns emerge about chip supply, export controls, and data jurisdiction.

2024–2028: AI as Critical Infrastructure

- AI is formally recognized as critical infrastructure in multiple major economies.
- GPU and accelerator scarcity becomes a persistent structural constraint.
- AI blocs begin forming around U.S., China, and emerging India–Gulf alliances.
- Data centers, undersea cables, and grid access become high-stakes geopolitical assets.

2028–2035: The Protectorate Pattern

- Nations unable to build sovereign compute lock into long-term hyperscaler contracts.

- Digital protectorates emerge: legally independent, digitally dependent.
- Nuclear reactors and SMRs for AI workloads are proposed and piloted worldwide.
- Cloud/security deals begin to resemble 21st-century mutual defense pacts.

2035–2045: The Megacenter Era

- Nuclear- and renewables-backed AI megacenters rise in energy-rich developing nations.
- AI systems shape culture, governance, labor markets, and national security doctrine.
- Most of the world is effectively governed by foreign AI platforms and policies.
- Data, energy, and compute corridors become as strategic as sea lanes once were.

2045–2050: Stabilization of the Dual Empire

- A dual structure solidifies: powerful states fused with global hyperscalers.
- Digital colonialism peaks: many countries operate as de facto digital protectorates.
- Attempts at digital non-alignment emerge but struggle against entrenched infrastructure.
- AI and nuclear power are treated as a single strategic domain in global negotiations.

Appendix C — The Five Pillars of Digital Sovereignty

Energy Sovereignty
Independent, abundant, and reliable power — increasingly including nuclear, hydro, and large-scale renewables — to support sovereign data centers and AI clusters.

Compute Sovereignty
National or allied GPU/accelerator clusters, sovereign clouds, secure networks, and at least partial domestic participation in chip design, packaging, or fabrication.

Data Sovereignty
Keeping sensitive national data (citizens, health, defense, finance, critical infrastructure) stored, processed, and governed under national law and independent regulatory oversight.

Talent Sovereignty
Training, attracting, and retaining elite talent in AI, chips, cybersecurity, systems engineering, and policy — and preventing a full brain drain into foreign hyperscalers.

Cultural Sovereignty
Building AI systems that understand and respect local languages, histories, and values, rather than importing default models aligned to foreign political and cultural norms.

Appendix D — List of Early Sovereign AI Initiatives

These initiatives mark early attempts by states to escape pure dependency and build sovereign or semi-sovereign AI capacity:

- **India** — IndiaAI Mission: national compute clusters, domestic models, sovereign stack initiatives.
- **France** — Mistral AI, national HPC tied to nuclear energy, European AI sovereignty strategies.
- **UAE** — Falcon models, G42 clusters, alignment with Western hyperscalers under sovereign branding.
- **Saudi Arabia** — $40B AI investment fund, plans for nuclear-backed compute and mega-projects (e.g., NEOM).
- **Japan** — National compute centers and semiconductor revival policies.
- **South Korea** — Sovereign models, alignment of AI strategy with world-class chip engineering capabilities.
- **Singapore** — Model governance leadership, sovereign cloud strategies, regional AI hub positioning.

Appendix E — Questions Governments Must Ask Before Adopting Foreign AI

1. Who owns the data once it is uploaded or processed?
2. Where, physically and legally, will the data be stored and backed up?
3. Who controls the models that interpret, categorize, and act on this data?
4. Can the nation continue operating critical services if cloud access is cut or throttled?
5. Are there credible local or allied alternatives today — or can they be built within 5–10 years?
6. How will dependency evolve over 10–20 years, as lock-ins, standards, and APIs harden?
7. Does the foreign AI embed foreign cultural, legal, or political values into local institutions?
8. Can the nation realistically build some degree of sovereign compute — alone or in alliance?
9. What exit strategies exist if the provider changes terms, is sanctioned, or is acquired?

These questions reveal hidden costs, long-term risks, and the degree to which "adoption" is actually a transfer of sovereignty

Appendix F — The Warning Signs of Becoming a Digital Protectorate

A nation is drifting toward digital protectorate status when:

- National ID, civil registry, or tax systems are fully outsourced to foreign clouds.
- Healthcare, biometric, or security data is stored and processed on foreign infrastructure.
- Military, intelligence, or border analytics rely on external AI services.
- The country depends on foreign GPUs/accelerators with no credible diversification plan.
- There is no national AI model or sovereign stack for public-interest use cases.
- There is no national or allied cloud infrastructure under genuine domestic control.
- There is no domestic chip strategy (design, packaging, fabs, or alliances).
- Data-sharing agreements are drafted primarily by the hyperscaler's legal team.
- Local universities and labs cannot train or fine-tune frontier-scale models.
- National strategies mention "AI transformation" but not energy, compute, or export controls.

If a nation checks more than four of these boxes, it is already on the path to digital protectorate status.

Appendix G — A Framework for Nations Seeking Digital Independence

Phase 1 (0–3 years): Foundation

- Build or secure sovereign data centers and trusted cloud regions.
- Establish a national AI & compute strategy council or task force.
- Invest heavily in STEM, AI, cybersecurity, and chip-adjacent education.
- Implement core data localization and security laws for critical sectors.
- Begin mapping current dependencies: where compute, data, and models live today.

Phase 2 (3–8 years): Expansion

- Deploy national GPU/accelerator clusters for research, public services, and startups.
- Develop sovereign or co-sovereign foundation models in major national languages.
- Expand local cloud infrastructure with strong public-interest mandates.
- Negotiate bilateral or regional alliances for energy, compute, and chip supply.

- Create funding mechanisms and incentives for domestic AI and chip startups.

Phase 3 (8–15 years): Maturity

- Invest in domestic chip R&D and, where feasible, fabrication or advanced packaging.
- Build nuclear-, hydro-, or large-scale renewables-backed AI megacenters under national control.
- Form or join regional AI alliances to pool compute and bargaining power.
- Establish cultural AI institutes to protect and evolve languages, histories, and narratives.
- Regularly stress-test national resilience to foreign cloud outages and export controls.

Nations that do not begin Phase 1 by ~2030 are unlikely to reach Phase 3 by ~2045, and will remain dependent on others' empires.

APPENDIX H — Sources & Further Reading

The following works provide deeper context, evidence, and analysis for readers interested in exploring the themes of digital colonialism, AI geopolitics, compute power, hyperscaler dominance, infrastructure imperialism, and the future of global sovereignty.

Arranged by topic for clarity.

1. AI, Compute Power & Technological Dominance

Reports & Policy Papers

- *Stanford AI Index Report* (Annual)
- *MIT Technology Review — Special Issues on AI, Compute & Energy*
- *McKinsey Global Institute — The Economic Potential of Generative AI*
- *OECD — AI Policy Observatory*
- *European Commission — High-Performance Computing Strategy Papers*
- *DARPA AI Next Campaign Papers*

- *U.S. Department of Commerce — CHIPS Act Implementation Reports*
- *International Energy Agency (IEA) — Data Centers & Electricity Forecasts*
- *World Bank — Digital Infrastructure & Cloud Trends*

Books

- Cade Metz — *Genius Makers*
- Amy Webb — *The Big Nine: How the Tech Titans & Their Thinking Machines Could Warp Humanity*
- Mo Gawdat — *Scary Smart*
- Nick Bostrom — *Superintelligence*

2. Digital Colonialism, Data Power & Cloud Dependency

Books & Academic Works

- Shoshana Zuboff — *The Age of Surveillance Capitalism*
- Nick Couldry & Ulises Mejias — *The Costs of Connection: How Data Colonizes Human Life*
- Michael Kwet — *Digital Colonialism* (papers & lectures)
- Parminder Jeet Singh — *Digital Industrialization* (South Centre Reports)

Journals & Articles

- *International Journal of Communication — Digital Colonialism Issue*
- *Global Data Justice Literature Journal*
- *Harvard International Review — Cloud Hegemony Articles*
- *Nature — Ethics of AI & Global Power Inequalities*

Think Tanks

- *Centre for International Governance Innovation (CIGI)*
- *Carnegie Endowment for International Peace — Digital Society Papers*
- *Chatham House — Cyber Power & Geopolitics*

3. Hyperscalers, Cloud Infrastructure & Platform Politics

Books

- Jonathan Taplin — *The End of Reality*
- James Bridle — *New Dark Age*
- Shoshana Zuboff — *Surveillance Capitalism*

Reports

- *Gartner Cloud Infrastructure Magic Quadrant*
- *Synergy Research Group — Hyperscaler Market Share*
- *Uptime Institute Data Center Industry Reports*

- *The Information — Cloud Power Investigations*

Analyses

- *Brookings Institution — Tech Platforms & Governance*
- *New America Foundation — Cloud Sovereignty Papers*
- *RAND Corporation — Strategic Value of Cloud to Government*

4. AI Geopolitics, Great Power Competition & Global Blocs

Strategic Reports

- *RAND Corporation — AI and the Future of Warfare*
- *Council on Foreign Relations — Tech Rivalry Papers*
- *NATO Cooperative Cyber Defense Centre Analyses*
- *Center for Security and Emerging Technology (CSET) —* Georgetown University
- *Hudson Institute — China and the New Tech Order*
- *Mercator Institute for China Studies (MERICS) —* China AI Strategy Papers

Books

- Henry Kissinger, Eric Schmidt — *The Age of AI*
- Kai-Fu Lee — *AI Superpowers*

- George Friedman — *The Storm Before the Calm* (geopolitical macro framework)

5. Energy, Nuclear Power & Data Center Megascale Infrastructure

Energy & Infrastructure Reports

- *International Atomic Energy Agency (IAEA)* — SMR & Microreactor Papers
- *International Energy Agency (IEA) — Tech Energy Outlook*
- *BP Statistical Review of Energy*
- *National Renewable Energy Laboratory — Data Center Cooling & Power Studies*

Books

- Vaclav Smil — *Energy and Civilization*
- Gretchen Bakke — *The Grid*
- Daniel Yergin — *The New Map: Energy, Climate, and the Clash of Nations*

Industry

- *Microsoft — Nuclear-Powered Data Centers Research Notes*
- *Google — 24/7 Carbon-Free Energy Studies*

6. Cybersecurity, Surveillance, Identity & Digital Rights

Books

- Bruce Schneier — *Data and Goliath*
- Edward Snowden — *Permanent Record*
- Ron Deibert — *Reset: Reclaiming the Internet for Civil Society*

Organizations

- Citizen Lab (Toronto) — Global Surveillance Reports
- Privacy International — Data & Human Rights Investigations
- EFF (Electronic Frontier Foundation) — Digital Rights Papers

Government & NGO Reports

- UN Human Rights Council — AI & Human Rights Reports
- OECD — Digital Identity & Cross-Border Data Flow Papers

7. Historical Perspectives on Empire, Power & Colonial Patterns

Books

- Shashi Tharoor — *Inglorious Empire*
- Sven Beckert — *Empire of Cotton*
- Niall Ferguson — *Empire*
- Benedict Anderson — *Imagined Communities*
- Yuval Noah Harari — *Sapiens* & *Homo Deus*

For parallels with corporate empires:

- William Dalrymple — *The Anarchy: The East India Company*
- Charles Maier — *Among Empires*

These works help contextualize modern digital power through historical patterns.

8. Culture, Language, and AI-Mediated Identity

Books

- Byung-Chul Han — *The Transparency Society*
- Mark Fisher — *Capitalist Realism*
- Rory Cellan-Jones — *Always On*

Reports

- UNESCO — AI and Cultural Diversity
- British Council — Digital Languages Studies
- MIT Media Lab — Algorithmic Culture Papers

9. Technical Resources for Deeper Understanding of Compute & AI

Technical Papers

- *Scaling Laws for Neural Language Models* (Kaplan, OpenAI)
- *DeepMind Technical Papers* on AlphaFold & large models
- *Nvidia Architecture Whitepapers* (Volta → Hopper → Blackwell)
- *Meta LLaMA Technical Reports*
- *Google's TPU Research Papers*

These papers inform the realities of compute scaling, model evolution, and hardware constraints.

10. Recommended Further Reading for Policy Leaders & Strategists

- *AI for Global Governance* — World Economic Forum

- *Computing and National Security* — U.S. National Academies
- *AI & The Future of Power* — Samir Saran
- *The Future of Cloud & Sovereignty* — Accenture Strategy
- *Global Risks Report* — World Economic Forum
- *UN Digital Cooperation Roadmap*

EPILOGUE — Standing at the Threshold

Every era of human history contains a moment when the existing order becomes untenable and a new one begins to form.
The agricultural revolution.
The rise of the first cities.
The birth of empires.
The industrial age.
The nuclear age.
Now, the AI age.

For decades, we believed digital progress meant liberation.
The internet gave people a voice.
Smartphones connected billions.
Information became abundant.

And yet — almost without noticing — we stepped into a world where the most powerful structures are not nations, armies, or political parties, but clusters of servers humming quietly in data centers most of us will never see.

The new empires have no borders.
Their reach is global.
Their language is code.
Their weapons are algorithms.
Their currency is data.
Their territory is the human mind.

But despite the scale and momentum of these forces, the story is not over.
The future has not hardened.
Nothing is inevitable.

This book has traced:

- the concentration of power in hyperscalers,
- the risk of digital colonialism,
- the formation of AI blocs,
- the dependency trap facing nations,
- the energy and compute demands reshaping geopolitics,
- the echoes of historical colonial systems,
- and the potential futures that still remain open.

The conclusion is stark but direct:

We stand at the threshold of two possible worlds.

One in which power centralizes into digital empires...
and one in which humanity uses technology to expand freedom, culture, and creativity.

The battle is not between humans and machines.
It is between systems that **concentrate** power and systems that **distribute** it.
Between digital colonialism and digital sovereignty.
Between inequality and possibility.
Between a controlled world and an open world.

The tools of this century are powerful.
The stakes are global.
The responsibility is ours.

We still have time to decide.
But not forever.

History has not yet chosen its path —
and neither have we.

About the Author

Basem Garada is a writer and systems thinker exploring how technology shapes human experience. With an interdisciplinary background spanning medicine, neuroscience, and complex systems, his work examines the long-term effects of artificial intelligence and digital infrastructure on perception, creativity, and everyday life.

The AI Empire reflects his interest in resilience, meaning, and how technological systems quietly reshape civilization from within. His writing approaches advanced technology through a human-centered lens, emphasizing reflection over prediction and clarity over abstraction.

Basem lives in the United States and continues to write about the future of humanity in an age of machine intelligence.

www.ingramcontent.com/pod-product-compliance
Lightning Source LLC
LaVergne TN
LVHW090337210926
848480LV00004B/1447
9798994956526